D1320115

Intelligence and Technology

The Impact of Tools on the Nature and Development of Human Abilities

THE EDUCATIONAL PSYCHOLOGY SERIES

Robert J. Sternberg and Wendy M. Williams, Series Editors

Marton/Booth • *Learning and Awareness*

Hacker/Dunlovsky/Graesser, Eds. • *Metacognition in Educational Theory and Practice*

Smith/Pourchot, Eds. • *Adult Learning and Development: Perspectives From Educational Psychology*

Sternberg/Williams, Eds. • *Intelligence, Instruction, and Assessment: Theory Into Practice*

Martinez • *Education as the Cultivation of Intelligence*

Torff/Sternberg, Eds. • *Understanding and Teaching the Intuitive Mind: Student and Teacher Learning*

Sternberg/Zhang, Eds. • *Perspectives on Cognitive, Learning, and Thinking Styles*

Ferrari, Ed. • *The Pursuit of Excellence Through Education*

Corno, Cronbach, Kupermintz, Lohman, Mandinach, Porteus, Albert/The Stanford Aptitude Seminar • *Remaking the Concept of Aptitude: Extending the Legacy of Richard E. Snow*

Dominowski • *Teaching Undergraduates*

Valdés • *Expanding Definitions of Giftedness: The Case of Young Interpreters From Immigrant Communities*

Shavinina/Ferrari, Eds. • *Beyond Knowledge: Non-Cognitive Aspects of Developing High Ability*

Dai/Sternberg, Eds. • *Motivation, Emotion, and Cognition: Integrative Perspectives on Intellectual Functioning and Development*

Sternberg/Preiss, Eds. • *Intelligence and Technology: The Impact of Tools on the Nature and Development of Human Abilities*

For a complete list of LEA titles, please contact Lawrence Erlbaum Associates, Publishers, at www.erlbaum.com.

Intelligence and Technology

The Impact of Tools on the Nature and Development of Human Abilities

Edited by

Robert J. Sternberg
Yale University

David D. Preiss
Yale University and Pontificia Universidad Católica de Chile

2005 LAWRENCE ERLBAUM ASSOCIATES, PUBLISHERS
Mahwah, New Jersey London

Senior Acquisitions Editor:	Naomi Silverman
Assistant Editor:	Erica Kica
Cover Design:	Kathryn Houghtaling Lacey
Textbook Production Manager:	Paul Smolenski
Full-Service Compositor:	TechBooks
Text and Cover Printer:	Hamilton Printing Company

This book was typeset in 10/12 pt. ITC New Baskerville, Bold, Italics.
The heads were typeset in ITC New Baskerville Bold.

Lawrence Erlbaum Associates, Inc., Publishers
10 Industrial Avenue
Mahwah, New Jersey 07430
www.erlbaum.com

Library of Congress Cataloging-in-Publication Data

Intelligence and technology : the impact of tools on the nature and development of human abilities / edited by Robert J. Sternberg, David D. Preiss.
p. cm.—(The educational psychology series)
Includes bibliographical references and indexes.
ISBN 0-8058-4927-0 (casebound : alk. paper)
1. Information technology—Social aspects. 2. Intellect. I. Sternberg, Robert J. II. Preiss, David, 1973– III. Series.

T58.5.I565 2005
153.9—dc22

2005004906

Books published by Lawrence Erlbaum Associates are printed on acid-free paper, and their bindings are chosen for strength and durability.

Printed in the United States of America
10 9 8 7 6 5 4 3 2 1

The editors dedicate this book to Jerome Bruner, whose pioneering work on the cognitive and developmental consequences of technology has opened new vistas in our understanding of human intelligence.

Contents

Foreword

This challenging collection of essays deals with the impact of evolving information technologies on human mental life and, indeed, on the nature and organization of human culture as a store of information-processing techniques. What topic could be more relevant to our swiftly changing contemporary world? For we are blessed, besotted, and threatened by such technologies and preoccupied by their uses. Some are seemingly benign, as when the Internet broadens the horizons of the young, or when computers take the Dickensian drudgery out of bookkeeping. Some are more worrying in their impact, as when one speculates whether information technology may promote imperialism by widening the gap between informationally adept military powers and word-of-mouth local insurgents. This volume is principally (though not exclusively) about the former, about changes in thinking, feeling, and relating to each other created by the current information revolution. But it goes beyond its influences on individual mental activity to consider how the new technologies might alter the cultures and the economies that come to rely on them.

A word about this last point first. It is hardly news that technological growth changes how life is lived on our globe. Is the new information revolution, like revolutions before it, going to skew still further the astonishing maldistribution of wealth on our planet? To paraphrase James Wolfensohn, the president of the World Bank, one sixth of the world's population owns 80% of the wealth, and another sixth subsists on less than a dollar a day. Indeed, worldwide wealth maldistribution is paralleled at national levels as well, with America as perhaps the most striking example. Two of the chapters in the pages following deal particularly with the puzzling psycho-cultural problems created by introducing "advanced" information technologies into cultures or subcultures where they are not ordinarily found, and both of them (one by Ashley Maynard, Kaveri Subrahmanyam, and Patricia Greenfield, the other by Michael Cole and Jan Derry) make plain that culture has a compelling effect in shaping minds and that technology makes huge differences.

The overwhelming consensus of this book's authors is that it is the *uses* of a technology that matters, that our minds appropriate ways of representing the world from using and relating to the codes or rules of available technology. Yet, it is not simply that we are shaped by the "tools" that we use, as in the by-now classic story of human evolution—that human beings are is not simply *Homo sapiens*, but also *Homo faber*. It is also that certain forms of tool use permit us to create a *metarepresentation* of the world and the uses of mind in coming to that representation. Technologies for storing, transforming, and appropriately retrieving information provide occasion for "turning around" on our own usages, for seeing them in new and more detached ways, and for sharing our representation with others. In the process, we enter and come to take for granted a world of knowledge that we can use.

This process cannot be oversimplified as *just* the amplification of human mental functions by computers—the amplification of memory by systems of storage-and-retrieval, of thought by problem-solving and trouble-shooting programs, etc. One soon learns, reading the pages of this intriguing book, that there is also something more than amplification involved, that there is something more "meta" about our approach to mental functioning that results. Indeed, under these circumstances the boundary between what is "inner" and what is "outer" gets more porous—as well it might. Is that crucial boundary threatened?

I must say one final thing to put this book into historical perspective. In 1956, I wrote a book along with two colleagues titled *A Study of Thinking*. It dealt with how people categorized the things and events of the world and their "strategies" for ordering their encounters with possible instances of the categories they were using. Historically minded commentators like to say now that the book helped precipitate what is now, in retrospect, called the Cognitive Revolution. At the time, it received a quite mixed reaction: Some felt, in those behavioristically inclined days, that it was too mentalistic. In fact, the book was inspired principally by John von Neumann's early work on computational theory, work largely unknown to psychologists in those days. It is interesting to note that it was only when computational theorists began discovering the programmatic nature of "machine problem solving" that the complementary relationship between human and machine problem solving came to be appreciated—and, indeed, when the possible importance of the latter as aids to the former became fully apparent.

Now, nearly a half-century later, we are all of us deep into the question of *how* human problem solvers use technologically proficient "free-standing," problem-solving programs as adjuncts in our efforts to get answers to our problems. Indeed, we are now asking how (not whether) formal problem-solving programs can help us formulate our problems better. This, in turn, (and not surprisingly) has led us to inquire whether human culture itself cannot be conceived of as a stored and shared collection of ways of formulating problems and, indeed, of storing customary approaches to their solution. To be sure, this is not altogether new: After all, it was a question that motivated both Plato and Aristotle. Indeed, Giambattista Vico was preoccupied with it some 3 centuries ago. They, too, wondered about how the human mind was affected by this "inner–outer" interaction.

The essays in this book now take it for granted, quite properly, that the mind is not locked away from the culture's treasury of formal problem-solving programs, that it uses them constantly and, indeed, constantly adds to the treasury. Now, finally, we are explicitly asking the crucial question of how this process affects the mind. As several of this volume's authors ask, how shall we now think of the relation between mind and the culture that both shapes and enables it? The final chapter pays a fitting tribute to the anthropologist Leslie White who, a half century ago, posed this problem in a particularly lucid and challenging way. I join the tribute. But I would like to broaden it to include the many others who have dared cope with this same problem, each in their own way—Vico, to be sure, but also Wilhelm Wundt, Emile Durkeim, George Herbert Mead, and, yes, Benjamin Lee Whorf and John von Neumann. All of them would read this book with wonder—and astonishment.

—Jerome Bruner
New York University

Preface

We live in a world that is increasingly dependent on technology. The diffusion of computers and information technologies has changed the nature of multiple activities that were previously accomplished using paper-based technologies. New technologies have not only been created by the cognitive skills of their inventors, these technologies have transformed the nature of cognitive skills. To illustrate, today the use of word processors is so prevalent that writing relates progressively less to the cultivation of expression on paper and more to effective computer use. There is evidence that this change restructures the writing process, as planning and reviewing with word processors involves more cognitive effort than does working in longhand (Kellogg & Mueller, 1993). Moreover, it is possible to correct errors and to restructure material in ways that were never possible without computer technology.

Computers and hand-held calculators have also changed the nature of mathematical abilities. In past times, for example, the number factor in Thurstone's (1938) theory was measured, in large part, by tests of arithmetic computation (Thurstone & Thurstone, 1941). Achievement tests would also emphasize arithmetic computation as one of two or three basic skills (with the others usually identified as arithmetic problem solving and perhaps arithmetic concepts). Today, such tests would seem to many people to represent an anachronism, as hand-held calculators and computers have rendered arithmetical–computational skills much less important than they were in the past.

Finally, there is evidence that computers have had an impact on the nature of visual skills as well. For example, there appears to be a causal relation between the amount of practice with computer applications (such as video games) and higher levels of performance in spatial and visual tasks (Greenfield, 1998; Subrahmanyam, Greenfield, Kraut, & Gross, 2001). Based on this evidence, Greenfield (1998) proposed that the proliferation of computer applications may be related to the reported increase in raw-score equivalents of nonverbal IQs during the last century (Flynn, 1987).

Certainly, the computer revolution has increased our awareness of the cognitive consequences of technology. The nature of its impact on cognition has been strongly debated among educators, however. The massive implementation of the use of computers in schools has actually raised the question of whether computers, by altering the technological base of literacy, are generating a new form of literacy (Tuman, 1992). Some researchers are optimistic about the prospects of using computers and new technologies in education to foster student learning (Reinking, 1998). Accordingly, some psychologists view computers as one of the main sources of change in the educational system (Gardner, 2000). Notwithstanding this optimistic stance, research has shown that computers do not improve academic achievement per se: Their impact depends on a positive confluence of several variables, such as student engagement, group participation, frequent interaction and feedback from mentors, and connections to real-world contexts (Roschelle, Pea, Hoadley, Gordin, & Means, 2000). Moreover, there is evidence that successful implementation of technology in the classroom is mediated by teachers' instructional philosophies (Becker, 1999). Thus, diffusion of computers in the schools not only has increased our appreciation of the cognitive consequences of technology but also of the social processes that mediate its impact.

There are other ways the computer revolution has called attention to the impact of technology on cognition, in addition to its impact on educational practices. As computer-based forms replace paper-based forms of communication, the average skills required in work settings have changed as well. The term *computer literacy* is commonly used to illustrate the fact that paper-related skills such as reading and writing are not enough to be a "productive citizen" in the information society (Reinking, 1998). However, the nature of these new skills is not completely clear. Consequently, the notion of computer literacy is used more as an intuitive notion than as a clear-cut construct. Additionally, the malleable nature of information technology has made it more difficult to delimit a definite notion of computer literacy.

In fact, there are no easily identifiable paradigmatic skills such as reading and writing in the computer arena. On the one hand, the skills composing computer literacy are technology-dependent, so they have changed as computer technology has evolved (Lin, 2000). For instance, programming was considered an important skill in the early 1980s, but less so today, when Internet-related skills seem to be more important. On the other hand, nowadays the skills necessary for being an empowered citizen not only require the mastering of a few computer applications—the technological-literacy factor—but also require the acquisition of the ability to use problem-solving intellectual capabilities in an information–technology context (Hunt, 1995; Lin, 2000). In brief, as software and computer applications evolve, they make perceptible the secularly changing nature of human abilities. The technological-literacy arena is consistent, thus, with Sternberg's claims that human skills are adaptive and, to a large extent, context dependent (Sternberg, 1985, 1990, 2000; Sternberg & Wagner, 1986).

An intimate relation between technology and cognition is not exclusive to the computer era, however. In effect, it has been suggested that writing should be given the status of a technology with the power to restructure human thinking (Olson, 1994). There is an essential relation between human intelligence and technology. Tools mediate the relation between mind and environment: At the same time that abilities are shaped by the technological setup that constitutes their environment, human beings actively create and use tools and devices to adjust and to shape their environment.

As a result, investigation of the cognitive consequences of technology should not be considered an auxiliary matter restricted to the inquiry on computer applications. In effect, studying the psychological side of cultural tools is key to understanding a number of relevant observable facts, such as the earlier-mentioned rise in IQ scores during the 20th century (Greenfield, 1998), the acquisition of language by the child (Tomasello, 1999, 2000), and the cognitive consequences of literacy (Olson, 1994), among others. Although these different phenomena may seem to belong to disparate areas of psychology, attention to technology could help us to understand the basic and generic mechanisms that underlie all of them. Consequently, the technology–cognition interface should receive more systematic attention and the integration between different psychological streams of research relevant to technology should be encouraged.

A long-standing tradition in psychology has been the systematic study of the role of tools in the development of higher psychological processes (Bruner & Olson, 1977; Luria, 1976; Vygotsky, 1978, 1986). In addition, within psychology, inquiries into the psychological consequences of technology is present in a wide range of its subdisciplines: cultural and social psychology, industrial/organizational and human factors psychology, and developmental and educational psychology. This volume presents a multifaceted, but unified, statement of the different approaches involved in the study of this relevant topic.

GOALS AND OVERVIEW OF THE BOOK

To our knowledge, there has been no previous book that puts together into one complete volume the progress that has been made in recent years across different perspectives. This volume is inclusive and multidisciplinary. It includes historical, comparative, sociocultural, cognitive, educational, industrial/organizational, and human factors approaches. Authors are researchers from different countries and varied research areas who have participated in order to stimulate international multidisciplinary discussion. This book can be useful to a wide audience interested in understanding the impact of technological tools on intellectual development. Moreover, it should foster dialogue between researchers and professionals from different subfields of the psychological sciences. Thus, it should be instrumental in unifying technological inquiries within psychology (Sternberg & Grigorenko, 2001; Sternberg, Grigorenko, & Kalmar, 2001).

The goal of the book is to have readers reflect on the impact of various technologies on human abilities, competencies, and expertise. Some of the questions addressed are:

- What is the impact of different technologies on human abilities?
- How does technology enhance or limit human intellectual functioning?
- What is the cognitive impact of complex technologies?
- What is the cognitive impact of the transfer of technologies?
- How can we design technologies that foster intellectual growth?
- How does technology mediate the impact of cultural variables on human intellectual functioning?

The diversity and richness of technology relates to different forms of abilities, competencies, and expertise. In consequence, many psychologists, educators, and others are interested in exploring the ways in which technology and human abilities interact, but they lack a handy source of information to satisfy their interest. We believe this volume provides them with relevant perspectives and information.

On a theoretical level, discussion regarding the interaction of technology and the human mind is instrumental in advancing our understanding of the role of cultural tools in the development of human intelligence. In an age that puts more and more emphasis on the biological basis of competencies or on the innate, long-time-ago-evolved capacities of the human mind, discussion of the interaction of technology and human abilities can play a balancing role in psychology.

With regard to this volume, authors were asked to address the following core questions:

1. What is the approach you use to study technology, tool use, and cognition?
2. Why do you use this approach?
3. What questions does your approach answer and not answer? What are its strengths and weaknesses?
4. How does your approach relate to other approaches?
5. What are the theory or theories underlying your work?
6. What data have you obtained based on your approach, and how do you interpret them?
7. Where do you see your research program going in the future?

Authors were instructed to write in a way that is interpretable to first-year graduate students in psychology, with no specialized background.

The foreword of this volume is written by Jerome Bruner, to whom the book is dedicated.

Part I deals with "Cognitive Technologies in Historical and Cultural Evolution." The chapters in this part deal with the history of cognitive technologies and how they have evolved with culture, but also, helped culture evolve.

Chapter 1, "Technology and Cognition Amplification," is by Raymond S. Nickerson. Technology, broadly conceived as the building of artifacts or procedures—tools—to help people accomplish their goals, predates recorded history. The practice of building tools that aid one or another human function—perceptual, motor, cognitive—is probably nearly as old as technology itself. The first part of this chapter focuses on the development of artifacts and procedures designed to facilitate calculating and computing: number systems, the abacus, logarithms, the slide rule, special-purpose devices, the general-purpose pocket calculator, and the modern electronic computer. In the second part, attention is turned to the question of what aspects of technology could benefit from amplification by current or near-future technology and the problem of determining whether any particular technological aid to cognition is doing more good than harm.

Chapter 2, "Technology and the Development of Intelligence: From the Loom to the Computer," is by Ashley E. Maynard, Kaveri Subrahmanyam, and Patricia M. Greenfield. This chapter elaborates on the work of Greenfield (1998). It argues that only an explanation of the worldwide rise in IQ that focuses on cultural history can account for the particular patterning of changes known as the Flynn effect. The strategy used to construct this argument goes as follows: (a) identify historical changes in the ecocultural niche that could account for these changes in test performance, (b) cite both traditional and "natural" experiments to demonstrate a causal link, and (c) develop theory and evidence regarding the mechanisms behind these causal links. Cultural bias in testing, language, intelligence, and adaptation are also discussed.

Chapter 3, "Technology and Intelligence in a Literate Society," is by David R. Olson. In exploring the well-known relation between literacy and intelligence, the traditional issues of nature–nurture, ability–achievement, and heritability all arise. This chapter addresses the nature of literate competence, why it is related to the generalized competence we call intelligence, and how both are related to the bureaucratic "rationalization" of society.

Part II is titled "Cognitive Consequences of Educational Technologies." It deals with how educational technologies affect the ways in which students and others think.

Chapter 4, "Do Technologies Make Us Smarter? Intellectual Amplification *With*, *Of*, and *Through* Technology," is by Gavriel Salomon and David Perkins. This chapter addresses the general questions of whether technology fundamentally changes anything in the mind's functioning beyond trivial changes of speed and ease of processing or accessibility to information, and whether it enables kinds of thinking that could not happen without it. Two ways of affecting minds are considered. One way pertains to effects *with* technology, attained through the partnership between technology and mind that enables the distribution or substitution of mental functioning, affords pertinent metaphors ("the mind as computer"), amplifies functioning past a tipping point, constituting new modes of individual or collective thinking, or redefines basic functions such as memory, imagery, and intelligence. The second way pertains to effects *of* technology, whereby exposure to it and usage of it may alter mental functions, leading to more or less lasting changes of, for example, mental representation, skills of processing, or conceptions of knowledge. Part of the chapter is devoted to the question of whether effects of technology should concern us in light of the ubiquity of technology: Why is it important if individuals become (say) better problem solvers if partnership with technologically based problem solvers are omnipresent? This leads to questions such as what role is left for "raw" ability and whether technology could possibly cultivate mental skills and not only lead to de-skilling.

Chapter 5 is "Cognitive Tools for the Mind: The Promises of Technology—Cognitive Amplifiers or Bionic Prosthetics?" by Susanne P. Lajoie. Technology has been touted as an educational change agent. However, the likelihood of such change occurring is greatly increased if cognitive theories guide the design of such technology. Current learning theories have become more inclusive in that cognition, motivation, and the social context in which learning takes place are considered as interconnected. Cultural and societal issues are considered in these new theories and, hence, new phrases such as *communities of learners* and *communities of practice* have arisen. This chapter examines the use of technology to promote these new conceptualizations of learning within specific educational communities of practice. The view of computer environments as cognitive tools for learning emphasizes the potential roles that computer-based learning environments (CBLEs) can play within classrooms (Lajoie, 2000; Lajoie & Derry, 1993). Only by introducing new cognitive learning tools into real classroom learning situations and studying how students learn through them can we know how to improve their design. The author's notion of computers as cognitive tools has always supported the position that technologies can be designed to amplify, extend, and enhance what learners know and understand. However, the author's former advisor, Richard E. Snow, once asked why she used the term *cognitive tool* rather than *prosthetic device* for the mind. This question is explored in the context of technology-rich learning environments that have been developed in the areas of medical problem solving, scientific reasoning, statistics, and cognition and instruction.

Part III is titled "Technological Partnerships at Work." It deals with technology in the world of work.

Chapter 6, "Work in Progress: Reinventing Intelligence for an Invented World," is by Alex Kirlik. Interaction with modern work environments is increasingly mediated by tools such as information technology and automation. As such, the scientific challenge of understanding tool use and the applied problem of tool design both require a detailed analysis of tools as mutually constrained by both their users and their environments of use. The use and design of a hammer, for example, points in two directions: The handle must be well adapted to the user's capacities for action, and the peen must be well adapted to the nails to be driven. The same observation applies to epistemic tools, such as computer displays, that mediate the relationship between the display user and the work environment. Not only must displays be user friendly in being easy to read and comprehend, they also must present an accurate representation of the remote, distal work environment that is the true target of adaptation (e.g., displays supporting aviation, process control, medicine, etc). This chapter presents a body of both basic and applied research focusing on this mediating role of technology in the modern workplace. This research draws heavily on psychological theory and method explicitly tailored toward articulating the proximal–distal relations (Tolman & Brunswik, 1935) that mediate interaction with the environment in both the information–judgment (inferential) and means–ends (action selection) realms.

Chapter 7, "Cooperation Between Human Cognition and Technology in Dynamic Situations," is by Jean-Michel Hoc. In dynamic situations where the human operator cannot control everything precisely (e.g., air traffic control, fighter aircraft piloting, etc.), quite autonomous machines are developed (such as the automatic aircraft conflict resolution device or flight management systems). Although such machines are not already very "intelligent" (in the sense of their adaptive power), humans adopt a cooperative attitude toward them. The framework of human–human cooperation is suitable to approach the human–machine relationships in this kind of situation. This chapter presents the results of research on human–human cooperation, as a model to approach human–machine cooperation and as a way to design computer support to human–human cooperation. The implementation and the evaluation of the framework for human–machine cooperation is described. The originality of the domain consists in its temporal constraints. Future developments will be delineated within cooperation with high speed processes like in-car driving assistance, where the human–machine relation is more often than not *subsymbolic,* cooperation taking place at the perceptual-motor control level.

Chapter 8, "Transferring Technologies to Developing Countries: A Cognitive and Cultural Approach," is by Carlos Díaz-Canepa. This chapter discusses the impact of transfer of technology on an organization. Incorporating new technologies is considered to be one of the most critical changes in the life cycle of an organization. In particular, the chapter proposes that the transfer of new technology involves a break in an organization's equilibrium. The achievement of a new equilibrium depends on a complex process of learning and adjustment, which happens in a context that is technologically, functionally, culturally, socially, physically, and economically diverse. Defining a set of criteria that help to manage technological change is instrumental to (a) delimiting the role played by both persons and the technological systems in the next state of the organization, (b) designing appropriate mechanisms of coordination, communication, and supervision, and (c) managing the relation between the different subsystems in the organization and its immediate environments.

Part IV is titled "Intelligent Technologies and Technological Intelligences." It deals with the interface between intelligence and technology.

Chapter 9 is titled "Technologies for Working Intelligence," and it is written by David D. Preiss and Robert J. Sternberg. This chapter advances a theory about the relation between technology and intelligence. The chapter proposes that there are three privileged ways for individuals to relate to technology. First, individuals invent new technologies to solve past and present practical problems. Second, individuals receive a technology as a part of their cultural heritage. Third, individuals adapt to technologies that are new innovations in their cultural background. These different paths to technology involve different processes of assimilation and accommodation. By inventing new technologies, the individual is shaping the environment, so his or her relation with the technology is more or less transparent. Technologies that are received by cultural transmission involve the transmission of *intentional affordances* (Tomasello, 1999) and, accordingly, shape human behavior. Finally, adaptation to new technologies involves a process of reciprocal adjustment between the technology and the individual. The authors propose that these three paths to technology represent "ideals." In effect, these paths cohabit so the relation between mind and technology is not a deterministic one. The chapter closes by suggesting that an intelligent (and wise) use of technology involves awareness of the modifiability of both mind and technology.

Chapter 10, "We Have Met Technology and It Is Us," is by Michael Cole and Jan Derry. As a general rule, both psychologists and laypersons alike adopt an overly narrow notion of what technology is, and of course there are long-standing debates about the meaning of the word *intelligence.* This chapter begins from a consideration of the concept of technology derived from the Greek concept. Any systematic form of activity involving specialized knowledge can be considered a technology. The particular aspect of technology, so understood, is the sedimentation of "systematic treatment" in artifacts, aspects of the material world that have been passed on over generations and that mediate goal-directed activity in the present. The sum total of artifacts possessed by the social group into which a child is born is used to designate the culture of the social group. From this perspective, there is an intimate relation between technology and intelligence (understood in its original meaning derived from Latin, and introduced into English in the 14th century), which is the ability to understand. Rather than being a property of the individual human brain, intelligence comes to be understood as constituted in culture and distributed among individuals, the artifacts they are able to mobilize, and the life tasks they confront.

REFERENCES

Becker, H. F. (1999). *Teaching, learning and computing 1998: A national study of schools and teachers: Internet use by teachers.* Retrieved from Sept. 17, 2004, http://www.crito.uci.edu/TLC/FINDINGS/internet-use/startpage.htm

Bruner, J. S., & Olson, D. R. (1977). Symbols and texts as tools of intellect. *Interchange, 8*(4), 1–15.

Flynn, J. R. (1987). Massive IQ gains in 14 nations: What IQ tests really measure. *Psychological Bulletin, 101*(2), 171–191.

Gardner, H. (2000). *The disciplined mind: Beyond facts and standardized tests, the K–12 education that every child deserves.* New York: Penguin Books.

Greenfield, P. M. (1998). The cultural evolution of IQ. In U. Neisser (Ed.), *The Rising Curve* (pp. 81–125). Washington, DC: American Psychological Association.

Hunt, E. B. (1995). *Will we be smart enough? A cognitive analysis of the coming workforce.* New York: Russell Sage Foundation.

Kellogg, R. T., & Mueller, S. (1993). Performance amplification and process restructuring in computer-based writing. *International Journal of Man–Machine Studies, 39*(1), 33–49.

Lajoie, S. (2000). *Computers as cognitive tools, volume two: No more walls.* Mahwah, NJ: Lawrence Erlbaum Associates.

Lajoie, S., & Derry, S. J. (1993). *Computers as cognitive tools.* Hillsdale, NJ: Lawrence Erlbaum Associates.

Lin, H. (2000). Fluency with information technology. *Government Information Quaterly, 17,* 69–76.

Luria, A. R. (1976). *Cognitive development, its cultural and social foundations.* Cambridge, MA: Harvard University Press.

Olson, D. R. (1994). *The world on paper: The conceptual and cognitive implications of writing and reading.* New York: Cambridge University Press.

Reinking, D. (1998). *Handbook of literacy and technology: Transformations in a post-typographic world.* Mahwah, NJ: Lawrence Erlbaum Associates.

Roschelle, J. M., Pea, R. D., Hoadley, C. M., Gordin, D. N., & Means, B. M. (2000). Changing how and what children learn in school with computer-based technologies. *Future of Children, 10*(2), 76–101.

Sternberg, R. J. (1985). *Beyond IQ: A triarchic theory of human intelligence.* New York: Cambridge University Press.

Sternberg, R. J. (1990). *Metaphors of mind: Conceptions of the nature of intelligence.* New York: Cambridge University Press.

Sternberg, R. J. (2000). *Practical intelligence in everyday life.* New York: Cambridge University Press.

Sternberg, R. J., & Grigorenko, E. L. (2001). Unified psychology. *American Psychologist, 56*(12), 1069–1079.

Sternberg, R. J., Grigorenko, E. L., & Kalmar, D. A. (2001). The role of theory in unified psychology. *Theoretical & Philosophical Psychology, 21*(2), 99–117.

Sternberg, R. J., & Wagner, R. K. (1986). *Practical intelligence: Nature and origins of competence in the everyday world.* New York: Cambridge University Press.

Subrahmanyam, K., Greenfield, P., Kraut, R., & Gross, E. (2001). The impact of computer use on children's and adolescents' development. *Journal of Applied Developmental Psychology, 22*(1), 7–30.

Thurstone, L. L. (1938). *Primary mental abilities.* Chicago: University of Chicago Press.

Thurstone, L. L., & Thurstone, T. (1941). Factorial studies of intelligence. *Psychometric Monographs No. 2,* 94.

Tolman, E., & Brunswik, E. (1935). The organism and the causal texture of the environment. *Psychological Review, 42,* 43–77.

Tomasello, M. (1999). *The cultural origins of human cognition.* Cambridge, MA: Harvard University Press.

Tomasello, M. (2000). Culture and cognitive development. *Current Directions in Psychological Science, 9*(2), 37–40.

Tuman, M. C. (1992). *World perfect: Literacy in the computer age.* London: Falmer Press.

Vygotsky, L. S. (1978). *Mind in society: The development of higher psychological processes.* Cambridge, MA: Harvard University Press.

Vygotsky, L. S. (1986). *Thought and language* (A. Kozulin, Trans.). Cambridge, MA: MIT Press.

List of Contributors

Jerome Bruner
School of Law
New York University
New York, New York

Michael Cole
Department of Psychology
University of California,
San Diego
La Jolla, California

Jan Derry
Institute of Education
University of London
London, United Kingdom

Carlos Díaz-Canepa
Escuela de Psicología
Pontificia Universidad
Católica de Chile
Santiago, Chile

Patricia M. Greenfield
Department of Psychology
University of California,
Los Angeles
Los Angeles, California

Jean-Michel Hoc
Centre National de la Recherche
Scientifique and Université
de Nantes
Institut de Recherche en
Communications et Cybernétique
de Nantes
Nantes, France

Alex Kirlik
Beckman Institute for Advanced
Science and Technology
University of Illinois
Urbana-Champaign, Illinois

Susanne P. Lajoie
Department of Educational and Counseling Psychology
McGill University
Montreal, Quebec, Canada

Ashley E. Maynard
Department of Psychology
University of Hawaii
Honolulu, Hawaii

Raymond S. Nickerson
Department of Psychology
Tufts University
Medford, Massachusetts

David R. Olson
University Professor Emeritus
Ontario Institute for Studies in Education
University of Toronto
Toronto, Ontario, Canada

David Perkins
Harvard Graduate School of Education
Harvard University
Cambridge, Massachusetts

David D. Preiss
Department of Psychology
PACE Center
Yale University
New Haven, Connecticut
and
Escuela de Psicología
Pontificia Universidad Católica de Chile
Santiago, Chile

Gavriel Salomon
Faculty of Education,
University of Haifa
Haifa, Israel

Robert J. Sternberg
Department of Psychology
PACE Center
Yale University
New Haven, Connecticut

Kaveri Subrahmanyam
Department of Child and Family Studies
California State University
Los Angeles, California

I

COGNITIVE TECHNOLOGIES IN HISTORICAL AND CULTURAL EVOLUTION

1

Technology and Cognition Amplification

Raymond S. Nickerson
Tufts University

Technology, broadly conceived as the building of artifacts or procedures—tools—to help people accomplish their goals, predates recorded history. As amplifiers of human capabilities, tools may be classified in terms of whether the abilities they amplify are motor, sensory, or cognitive in nature. Those that amplify motor capabilities (muscle power, carrying capacity, striking force, etc.) include lever, shovel, hammer, wheel, and countless modern machines that embody the same principles. Those that amplify sensory capabilities include eyeglasses, microscopes, telescopes, audio amplifiers, and modern instruments that extend detection of both electromagnetic radiation and sound waves far beyond the range that can be sensed by the unaided eye or ear. Tools that amplify cognition include symbol systems for representing entities, quantities, and relationships, as well as devices and procedures for measuring, computing, inferencing, and remembering.

The boundary between tools that aid cognition and those in the other categories is not sharp. Tools that amplify sensory capabilities, for example, greatly extend our ability to observe the world of the very small and the very large (and very far away) and, in doing so, enrich our cognitive understanding of the universe immeasurably. Tools that extend muscle power or that make possible the manipulation and precise control of devices too small to be constructed or controlled by hand also are essential to the modern scientific enterprise and therefore also help amplify cognition in a very real, if indirect, way. Nevertheless, the distinction among tools that extend sensory capability, those that extend muscle activity, and those that amplify cognition is conceptually meaningful, even if sharp dividing lines cannot be drawn.

It is easy to overlook or take for granted cognitive technologies that we use to great advantage in everyday life. Consider, for example, alphabetization. How the order of the letters of the alphabet (of English or any other alphabetic language) became established is not clear, but the fact that a fixed order did become established and that it is universally recognized makes the alphabet an invaluable tool for organizing and finding information in dictionaries, directories, encyclopedias, indexes, almanacs, atlases, and so on. Compare the problem of finding a bird in a bird book if one knows its name with that of finding the bird if one knows only what it looks like. The first task is easy because of the alphabetic organization of the book's index; the second is difficult because there is no comparably simple way of organizing information on the basis of visual features. The ease with which the importance of such a powerful tool is overlooked is illustrated by White's (1962) observation that an elaborate article on "Alphabet" in the *Encyclopedia Britannica* failed to mention its own organizational system. "That we have neglected thus completely the effort to understand so fundamental an invention should give us humility whenever we try to think about larger aspects of technology" (p. 488).

Much cognitive technology has developed in more or less the same way that the ability to stand, to walk, to run develops in the child. People have done what comes naturally as they have tried to extend their abilities to cope with the problems and to respond to the challenges that life presents. One could make a very long list of technological inventions of antiquity that have served to aid cognition. Here, I want to begin with a focus on some old technological developments that increased people's ability to think quantitatively—to count, measure, and compute—and then to consider how the invention of the modern digital computer and associated developments have extended the range of possibilities for the amplification of cognition.

THE DEVELOPMENT OF MATHEMATICAL SYMBOL SYSTEMS

The origin of symbol systems for representing quantities is not known with certainty, but according to one theory, both writing and number representation developed from the use, more than 11 millennia ago, of token systems to represent the number of items in a collection (Schmandt-Besserat, 1978). The tokens were small objects, coded perhaps by shape, size, and markings and were probably used as bills of lading for merchandise that was being transported from place to place. Initially, the tokens were enclosed in sealed containers, which were broken open by recipients of shipments in order to verify that all that had been shipped had reached the intended destination. Over time, merchants began representing what the sealed containers contained by marks on the containers, and eventually it became apparent that such marks could serve the same purpose as the enclosed tokens, so the tokens were no longer used.

After these murky beginnings, distinctly different symbol systems were developed by several cultures (Ifrah, 1987; Menninger, 1969). The Hindu-Arabic system that appeared around the 7th or 8th century and is now used almost universally is based on several principles—one-for-one mapping, the use of a standard quantity to serve as a base or radix, one-for-many substitution, use of a single symbol to represent a multiple quantity (e.g., 3 instead of, say, 111), the use of a symbol for representing an empty set (0), and the use of position to carry information. Predecessors of this system also made use of some of these principles—apparently some were invented or discovered more than once independently—but none of them made use of all of them. The current system is more abstract in some ways than most of its predecessors, but it is extremely versatile (faciliating the expresssion of an unlimited range of numbers) and greatly simplifies computation (Ifrah, 1987; Nickerson, 1988). Jourdain (1913/1956) considered the Hindu-Arabic system to be responsible for making many arithmetical problems that were formidable challenges to the ancients seem easy to us.

Archimedes (3rd century B.C.) is widely considered to have been one of the greatest mathematicians who ever lived, but he was seriously limited in what he could do by the Greek system for representing numbers, which was not conducive to computation. Gauss regarded the fact that Archimedes failed to invent a place notation system for numbers as "the greatest calamity in the history of science" (Bell, 1937, p. 256). It was a calamity, in Gauss's view, because he believed that if Archimedes had discovered the place notation principle, which he thought to be within Archimedes's capability to do, many of the subsequent accomplishments in mathematics and science might have occurred centuries before they did.

The history of the development of mathematical notation, especially that of symbolic algebra, also illustrates the importance of a good representational system as an aid to thought. The Greeks had a good system for representing geometric relationships, which tend to be static, but not for representing relationships among variable quantities, which are dynamic. Maor (1994) argues that the inadequacy of the Greeks' system for representing algebra helps explain why they did not discover calculus, despite the fact that Archimedes managed to apply Eudoxus's "method of exhaustion," which came close to modern integral calculus, to the finding of the area of the parabola.

LOGARITHMS AND SLIDE RULES

Tools can amplify cognition either by facilitating reasoning directly or by reducing the demand that the solution of a problem makes on one's cognitive resources, thereby freeing those resources up for other uses. The latter types of tools might be considered labor-saving tools in the cognitive domain—thought-saving tools if you will. A 19th-century instruction manual for the carpenter's slide rule (similar in principle to the engineer's slide rule that was, until fairly recently, the engineer's constant companion) describes what the slide rule offers to the user in such terms:

> The labour and fatigue of manipulating long series of figures for nautical and astronomical purposes had long been felt to be irksome to those engaged in it . . . One of the earliest attempts, however, by mechanical means, to lessen and facilitate this labour, was made more than two hundred and fifty years ago by Baron Napier, of Merchiston, in Scotland; and this attempt was the precursor of the Sliding Rule. (Anonymous, 1880)

The 16th-century Scottish mathematician, John Napier, of whom the author of this manual spoke, is remembered primarily for his invention of logarithms. Napier had the insight that if numbers are expressed in exponential form, multiplication and division of different powers of the same number can be accomplished by addition and subtraction, respectively, of exponents: thus

$$(3^3 \times 3^4 \times 3^7) \; (3^5 \times 3^2) = 3^{(3+4+7-5-2)} = 3^7.$$

Things are easy as long as the numbers involved are integral powers of the same number (the base, also sometimes called the radix or root), but they get complicated quickly when one wants to deal with numbers that do not meet this criterion. Napier spent 20 years working out logarithms that, for the most part, were not integers and published tables of his results. Napier also invented a system of rods that, because they were made of bone or ivory, became known as "Napier's bones," by which multiplication and division could be done, albeit in a somewhat tedious way.

In 1620, Edmund Gunter, an English mathematician, designed a rule on which the numbers were spaced in such a way that their distances from the end of the rule were proportional not to the numbers themselves but to their logarithms. With the use of such a rule and a pair of compasses, problems of multiplication and division were reduced to those of addition and subtraction. To multiply 3 by 4 with this rule, for example, one would first set the compasses by placing one leg on 1 and the other on 3 and then adding the resulting distance to 4 by setting one leg on 4 and seeing where the other leg would land. Because of the logarithmic spacing of the numbers on the rule, the distance between 1 and 3 represented the logarithm of 3 and when this distance was added to 4 (whose distance from 1 represented the logarithm of 4), the result would be the number (12) whose distance from 1 represented the logarithm of the product of 3 and 4. The same procedure could be used to multiply 30 and 40 or .3 and .4 or any other numbers, and division was accomplished by the inverse procedure.

Gunter's rule, especially with the inclusion of logarithms of trigonometric functions, simplified calculations of the sort required in navigation considerably. Nevertheless the use of compasses in calculating was tiresome and errors could be made easily. Someone noticed that the compasses could be dispensed with and computation would be easier and less error prone if one simply laid two Gunter's scales side by side. To multiply one number, x, by another, y, for example, one simply placed the number 1 on one scale beside x on the other and then read from the latter scale the number opposite y on the former.

About 10 years after Gunter invented his rule, someone got the idea of adjoining two such rules so one could be slid along the other in a controlled fashion and the logarithmic slide rule was born. There is some debate as to who first had this insight. Edmund Wingate and William Oughtred are generally considered the primary candidates. The anonymous writer of the 1880 instruction manual mentioned previously credits the idea of the sliding rule to William Forster, a pupil of Oughtred's. Apparently the idea was more easily conceived than executed; a functional sliding rule was not built until some time, perhaps decades, following its invention. Again, the author of the instruction manual writes:

> It may be supposed that at first the sliding rule was not much used, if only from the difficulties found in its construction. This may be judged of somewhat from the following extract from the interesting diary of Mr. Pepys, secretary to the Admirality in the time of Charles II. Under the date of the 10th of August, 1664, Pepys says: 'Abroad to find out one to engrave my tables upon my new sliding rule with silver plates, it being so small that Browne, that made it, cannot get one to do it. So I got Crocker, the famous writing master, to do it, and I set an hour beside him to see him design it all; and strange it is to see him, with his natural eyes to cut so small at his first designing it and read it all over, without missing, when, for my life, I could not, with my best skill, read one word or letter of it.' (Anonymous, 1880, p. 11)

The same writer points out that the difficulty of obtaining good rules was not much less nearly a century later. Only an exceptionally skilled craftsman could produce scales with the degree of precision needed to ensure accurate calculations with a rule. One notable craftsman who was up to the task was James Watt, who is remembered primarily for turning Thomas

Newcomen's steam engine into a viable commercial product; but the necessity of making the rules by hand and the limited number of people who could do so meant that the devices were not readily available to people of modest means.

Numerous improvements in the designs of slide rules were made during the 17th and 18th centuries and many different scales were invented to make the instruments useful for a variety of purposes, including gauging, ullaging, and the computing of taxes and tariffs. Among the names commonly associated with such developments are those of Robert Bissaker, Henry Coggeshall, Thomas Everard, William Nicholson, John Robertson, and Robert Shirtcliffe.

About 1811, Joshua Routledge, an English engineer, designed a rule that, in addition to Gunter's logarithmic scale, contained scales with several *gauge points*, or constants that facilitated certain calculations of special interest to engineers. This rule became known as the engineer's slide rule. The carpenter's slide rule, which was designed a few decades later by Sir William Armstrong, a British hydraulic engineer, was very similar to Routledge's rule but differed in certain respects that made it more convenient for doing the types of calculations needed in carpentry (Roberts, 1982). Instruction manuals for the engineer's rule (Routledge, 1867) and the carpenter's rule (Anonymous, 1880) were published by John Rabone and Sons, the first company established to manufacture rules. Reprints of both were published by the Ken Roberts Publishing Company in 1982 and 1983.

Calculating with a slide rule requires judging how the scale marks on a slider are aligned with those on one of the scales of the stationary part of the rule. Especially with the earlier rules, the precision with which this could be done was quite limited. At some point in the 18th century, someone—perhaps John Robertson (Hopp, 1999)—got the idea of increasing the accuracy of reading by adding a sliding *index, runner,* or *cursor,* a transparent device with a cross hair, that could be positioned over the area of the rule where the alignment had to be read. The runner was especially helpful in comparing readings on noncontiguous scales and was even more helpful when made of magnifying glass. (Although, strangely, the runner did not become a standard part of the majority of rules until early in the 20th century.)

Slide rules were made by hand until about the middle of the 19th century, when mechanized techniques began to be used for the purpose. The mid-19th century also marks the design by Amédéé Mannheim of the rule that was to remain more or less the design of preference for the next century. The combination of the much finer accuracy with which scales could be marked by machinery, the increasing addition by manufacturers of the runner, the beginning of the use of celluloid in some cases instead of wood, and the lower cost of manufacture greatly increased the usefulness of slide rules and their general availability to prospective users. Although the most common slide rules were straight, many circular ones were also made. Superficially, the circular rules appear to operate differently than do the straight rules, but the underlying principles are the same.

The slide rule became very popular and was used for many purposes, especially in Europe, during the 19th century. By the early part of the 20th century, it had become an indispensable tool for engineers. Facility with this device was the *sine qua non* of competence of any member of a profession for which the performance of nontrivial calculations was an important aspect of the work. Cajori (1910/1994), who gives an extensive account of the invention, refinement, and uses of the logarithmic slide rule, lists 256 different rules that were made between 1800 and the time of his writing in 1910. These include "rules designed for special kinds of computation, such as the change from one system of money, weight, or other measure, to another system, or the computation of annuities, the strength of gear, flow of water, various powers and roots. There are stadia rules, shaft, beam, and girder scales, pump scales, photo-exposure scales, etc." (p. 73). New scales for rules were constantly being invented for special purposes and designers found ways to include many scales—in some instances more than 30 (Hopp, 1999)—on a single rule.

Hopp (1999) lists several hundred major makers and retailers of slide rules that operated during the 20th century. Nearly all of them had stopped manufacturing or selling slide rules by the end of the third quarter of the century. What was a booming business at the beginning of the century was dead at the end of it, by which time slide rules had been largely replaced by pocket calculators, which could be used to calculate anything that could be calculated with a slide rule and much else besides. Calculators gave precise answers to problems for whch slide rules gave only approximations, and they could be used to good effect with considerably less training than was needed to master the use of a slide rule.

SPECIAL-PURPOSE SLIDE RULES AND RELATED DEVICES

The rules discussed in the preceding section all derived from Gunter's logarithmically spaced scale, although many of them contained other scales as well. Slide rules have also been designed for purposes that do not require the use of logarithms and that therefore do not contain logarithmic scales. Such rules typically are useful only for a specific purpose or category of problems.

Many companies distributed complimentary rules for advertising purposes, especially during the first half of the 20th century. Often such rules contained, in addition to company advertisements and conventional inch and fractional-inch scales, information that would be assistive for projects that made use of the company's products. Sometimes the assistive information was in the form of one or more special-purpose scales. I have a small collection of such devices, mostly old, and will mention several by way of illustrating the range of functions they served.

- A circular celluloid device distributed by Sunkist for calculating the costs and selling prices of oranges (lemons on the flip side), per dozen, given the cost per box and the number of oranges (or lemons) per box, and assuming a specified mark-up.
- A similar device produced by Post's Cereals, for computing the per-package retail sale price of a product, given the wholesale price of a box of packages, the number of packages per box, and the desired profit percentage.
- A device (copyrighted in 1924) advertising Mead's dextri-maltose, a dietary supplement for babies, that permits one to find the recommended mixture of milk, water, and dextri-maltose for each feeding and number of feedings in 24 hours, given the baby's age and weight.
- A shop-cost calculator produced by General Electric (reflecting costs in 1953) that calculates the labor costs for operating a shop, given an hourly wage, number of operations performed per minute, and an overhead rate.
- A device distributed by the Esso Corporation (predecessor of Exxon) that can be used to calculate distance traveled by an airplane, given speed and time in flight; gallons of fuel consumed, given gallons consumed per hour and time in flight (or fuel consumed per hour, given total consumed and time in flight); and speed, given distance traveled and time in flight. It allows for making a correction in air speed, given the temperature and altitude, and can compute drift angle and ground speed, given course heading, wind velocity and direction, and air speed.

Other devices in my collection compute: (a) the appropriate torque setting on an adjustable torque wrench for a wrench extension of a given length, (b) a correction factor for a steam flow system, given a calibrated pressure and an operating pressure, (c) the feet rate for a turning, boring, or milling machine, given certain parameters of the machine, stock, and desired product, (d) relative humidity, given dry-bulb and wet-bulb temperatures, (e) certain motor data, given motor horsepower, and (f) conduit data, given wire size and composition (copper or aluminum).

The idea that people build cognitive functionality into tools and thereby relieve themselves of the some of the cognitive burden that certain tasks would impose in the absence of these tools has some currency among psychologists (Salomon, 1993). The principle is well illustrated by the devices just described and many others that could be mentioned. Many of the computations performed by the devices are quite simple and could be done with pencil and paper if one knew the appropriate formulas. Nevertheless, such devices relieve their users of the need to do the computations, which can become tedious if they have to be done frequently.

NON-SLIDE INFERENCING RULES

The history of the invention and refinement of the slide rule, in its various instantiations, and of the numerous innovations that extended its usefulness into countless domains of activity has many examples of insight and ingenuity. It is a remarkable device and has served well the purpose of facilitating problem solving in many contexts by simplifying computations that, without its use, could be cognitively burdensome. Insight and ingenuity are also seen in the design of many other rules that do not use a slider, but that also simplify problem solving, often by transforming what, without the use of the rule, would require computation or some form of inferential reasoning into a scale-reading task. Examples include gauging, wantage and ullage rules, lumber rules, and shrink rules.

Gauging rods were used to determine the capacity of casks. Wantage rods were used to determine how much liquid was "wanting" from a cask, which is to say by how much the cask lacked being full. (*Wantage* is an American term; in Great Britain the term is *ullage* and what is measured by a ullage rod is the contents rather than what is missing.)

Because casks were made entirely by hand, each was, to some degree, unique and its capacity, as determined with the use of a guaging rod, was an approximation. Several mathematical formulas were used to express the theoretical capacity (volume) of a cask as a function of a few parameters: its head diameter (if its two heads differed, an average of the two was taken), its belly (greatest) diameter, and its length. Pachham (1997) gives six different formulas, which give six different, but relatively close, values of volume for the same set of parameters.

Gauging rods were calibrated in such a way that one could read off the capacity from a single measure (the diagonal from the bung hole to the groove for the lid on the opposite side of the bung hole, or the average of two such measures taken from the top and bottom lid) and some other known parameter of the cask. Scaling of gauging rods was possible because the capacities of two casks of the same shape but different size had been shown to be related as the ratio of the cubes of their diagonals.

Wantage rods typically had several, often as many as eight, scales, each calibrated for a specific size of cask. For example, I am now looking at a wantage rod that has scales for casks with capacities of 16, 32, 64, 84, 110, 130, 140, and 150 gallons. Some wantage rods were calibrated to be inserted into the top of a vertically standing cask, others calibrated to be inserted through the bung hole of a cask lying on its side. (*Cask* appears to be a more generic term than *barrel*. The latter referred originally to a cask of a particular size—36 imperial, or 31.5 U.S., gallons [Pachham, 1997]. Over time barrel acquired an increasingly inclusive connotation until it became more or less synonymous with cask.)

The purpose of measuring instruments is to provide information to their users. Numerous types of instruments and scales were developed that not only measured but, in effect, performed some calculation, thus relieving the user of the need to do it. Gauging and wantage rules are cases in point. So are the lumberman's timber rules that permit the reading of board feet from measures of timber diameters, the foundryman's shrink rules that are scaled to accommodate shrinkage in castings when they cool, and many other examples that could be given.

OTHER "PRE-COMPUTER" COMPUTING DEVICES

To most of us today, the word *computer* means a particular high-speed electronic device, but devices to perform or aid computing and calculating probably predate recorded history. No one knows when people started counting and calculating. Presumably counting is an older capability than calculating, but it is hard to imagine that the ability to tally the items in a set could have existed very long before someone noticed that a set containing, say, seven items was equivalent in number to the combination of one containing five and another containing two. However, and whenever, the abilities to count and to calculate originated, it appears that the use of symbols and other artifacts to assist the process of counting, or to record the results of doing so, predates the development of written language by many millennia. I have already mentioned Schmandt-Besserat's (1978) theory of how tokens and then symbols were used to represent bills of lading as early as the 9th millennium B.C. There is also at least suggestive evidence that notched bones were used to record phases of the moon, number of animals killed by hunters, and other matters of interest as many as 20,000 to 30,000 years ago (Ifrah, 1987).

Various ingenious mechanical devices that assist in counting and record keeping have been invented and used by different cultures over the centuries. The abacus, which was probably invented independently more than once, was widely used many centuries ago (in some form) in Egypt, Greece, China, Japan, and Russia; it embodied some of the same principles of representation as does the Hindu-Arabic system, as previously noted. As the practical applications to which mathematics is put have steadily increased, numerous efforts to build mechanical devices that could aid the making of computations have also been made, some of which have yielded products that were very helpful to some set of users, at least for a limited time.

The French philosopher–mathematician, Blaise Pascal, built a mechanical calculator, completing it in 1642, that could add, subtract, multiply, and divide. To accomplish the carry for addition, Pascal invented the technique that is still in use in many mechanical counters, such as odometers. He concatenated a series of discs, each numbered from 0 to 9, in such a way that when a disc was moved from 9 to 0, a ratchet caused the disc to its left to be advanced one digit. Pascal's device accomplished multiplication and division by repetitive addition and subtraction. About 30 years after Pascal completed his calculator, Gottfried Wilhelm Leibnitz succeeded in building a device that could multiply and divide directly.

Undoubtedly the name that is most strongly associated with devices that led directly to the development of the modern electronic digital computer is that of the eccentric but prescient English mathematician, Charles Babbage. Babbage made several inventions that had nothing to do with computing, but his main interest through most of his adult life was the possibility of building a mechanical calculator that would be capable of automatically performing the kinds of computations required for generating the tables of logarithms and other complicated functions that were then done by hand and often contained numerous errors. He planned and tried to build first a machine that he called the Difference Engine and then one he called the Analytic Engine during the period from 1822 to 1833, or thereabout. The machines were never fully completed, and the effort became known among his colleagues as Babbage's folly; however, many of his ideas have stood the test of time and are reflected in the design of modern computers. At some point, his work attracted the attention of Ada Augusta, Lady of Lovelace (for whom the computer language Ada was named), who apparently was better able to see the promise in Babbage's work than were his professional colleagues. She was responsible for documenting the work and has been credited with being the first to propose the idea of storing a program within the machine.

Babbage's Analytic Engine was to be composed of three parts: a store, a mill, and a control. Actually, he finally got around to the idea of two stores—one to hold data and one to hold instructions. The mill, which was analogous to the processing unit in modern computers, was the component that worked on data. The function of the control unit was to bring data out of the store so that they could be operated on by the mill, and to control the sequencing of events. Input and output components were not emphasized by Babbage, perhaps because in the machine that he designed, data were entered by hand and the output was simply the result of an arithmetic computation.

THE ELECTRONIC DIGITAL COMPUTER

Although Babbage's ideas presaged its development in a variety of ways, the modern programmable electronic digital computer did not appear on the scene until more than a century after he laid out his plans for the Analytical Engine. Babbage's vision could not easily be realized with a mechanical device; implementation of a machine that could do what Babbage had in mind had to wait on the development of the technology of electronics. By about the middle of the 20th century, the technology had developed to the point that the building of machines capable of doing complicated computations under the direction of stored programs became feasible. Events that helped mark the beginning of this era include completion of Howard Aiken's Mark I by IBM in 1944, completion of J. Presper Eckert and John Mauchly's ENIAC with its 17,000 vacuum tubes in 1946, invention of the transistor by John Bardeen, Walter Brattain, and William Shockley in 1948, and the first commercial sales of electronic digital computers about mid-century.

Even before computers became widely used for straightforward data processing, their potential for amplifying human cognition (intellect, reasoning, problem solving) was promoted by a number of visionaries, notably Vannevar Bush (1945), Ross Ashby (1956), J. C. R. Licklider (1960), Simon Ramo (1961), and Douglas Engelbart (1963). By the early 1960s sufficient interest had been shown in the subject to motivate the convening of a symposium on "Computer Augmentation of Human Reasoning" by the Bunker-Ramo Corporation and the Office of Naval Research. The symposium was held in Washington, DC, in 1964, and the proceedings were published the following year (Sass & Wilkinson, 1965).

The augmentation envisioned was extensive. Engelbart (1963) defined it as increased capability to approach, comprehend, and solve complex problems.

> Increased capability in this respect is taken to mean a mixture of the following: that comprehension can be gained more quickly; that better comprehension can be gained; that a useful degree of comprehension can be gained where previously the situation was too complex; that solutions can be produced more quickly; that better solutions can be produced; that solutions can be found where previously the human could find none. And by "complex situations" we include the professional problems of diplomats, executives, social scientists, life scientists, physical scientists, attorneys, designers—whether the problem situation exists for twenty minutes or twenty years. (p. 1)

Engelbart noted that the term "intelligence amplification" from Ashby (1956) is an appropriate descriptor of the goal of augmenting human intellect inasmuch as "the entity to be produced will exhibit more of what can be called intelligence than an unaided human could demonstrate" (p. 10).

COGNITION AMPLIFICATION BY INFORMATION TECHNOLOGY

By information technology I mean computer and communication technology, in combination. This technology has the potential of aiding cognition in the ways envisioned by Engelbart and in numerous other ways as well, some of which are beginning, but only beginning, to be realized. In what follows, I mention a few of the possibilities that strike me as interesting, especially in view of evidences in the psychological literature regarding aspects of human cognition that could benefit from amplification.

Among the more hopeful uses of information technology to amplify cognition are the many possibilities and prospects of providing aids for people with sensory or motor impairments that are impediments to their realization of their full cognitive potential. Reading machines for blind people, cochlear implants for people who are deaf, and computer-based communication devices for people who have severely limited speech are noteworthy examples of past accomplishments in this area. But much remains to be done and work is progressing on many fronts, including the possibility of using various physiological signals (electrical signals from muscles, corneal–retinal potentials, electrical signals from the brain) as inputs to computer-based systems that would give severely physically handicapped individuals much greater control over their immediate environments (Lusted & Knapp, 1996; Nicolelis & Chapin, 2002).

Here I do not focus on cognition aids designed explicitly for people with one or another type of disability, but many of the possibilities for amplifying cognition generally have the potential, I believe, to be of special interest and use to people with disabilities, and this is a considerable plus. In what follows, I mention a few of the many opportunities for the application of information technology to the amplification of cognition. In some cases, something can be said about progress that has already been made.

Information Finding

Many tools have been developed to help people find information, including encyclopedias, dictionaries, atlases, directories, almanacs, and other reference books. Information that is distributed in newspapers and periodicals is typically organized to facilitate the finding of items of interest. Books that are intended to inform generally have a table of contents and one or more indexes. But despite the various methods that have been devised to help people find their ways to information they need or want, probably most of us still spend more time than we would like searching for information that proves to be hard to find.

I have argued elsewhere that information technology has already had a major impact on cognition by making information far more accessible than it has ever been before and that its potential for increasing information accessibility further is enormous (Nickerson, 1986, 1995). So far, the promise has been realized, however, to only a modest degree. Search engines—Google, Yahoo, and others—manage to search an amazingly large body of information in a very short time and can be extremely useful. However, what is now possible with these types of tools is just enough to fire one's imagination regarding what may be possible in the foreseeable future. The engines that currently exist, as impressive as they are, lack intelligence and, for that reason, are limited in their ability to tailor a search to one's interests, as would a human expert.

A computer network that simply facilitates communication among human beings to other human beings turns out to be a very effective information-finding tool. I am on a mailing list that contains a few hundred members. Almost every day there are several messages from people on the list—directed not to anybody in particular, but to the list as a whole. Major topics of comments change every few days. "Does anyone know ____?" appears often. Almost always, the inquirer gets an answer, typically many. A few years ago, I analyzed a set of about 1,000 consecutive postings on a company electronic bulletin board (Nickerson, 1994). About one third of all messages posted were either "Does-anyone-know" type questions (21%), replies to these questions (6%), or notices of desires to buy, rent, borrow, or otherwise obtain something (7%). These data undoubtedly grossly underestimate the number of replies that "Does-anyone-know" questions evoked, inasmuch as askers often requested that replies be sent directly to them and not posted to the bulletin board, and they sometimes posted summaries of the replies received after a few days.

Requesting information from a community of computer users is different, of course, from requesting information from a computer's own information store. A question for continuing research is that of how a computer system that has access to an encyclopedic information store can be made to respond more like a human expert in certain ways. I say "in certain ways" because one wants the computer to respond like a human expert in some ways but not in others. One wants it to be like a human expert in having extensive knowledge of the area of interest, being able to communicate in natural language, being able to deliver the information desired without a lot of irrelevant material, and being forthcoming with helpful suggestions. But one wants it to be unlike a human expert in having extensive knowledge of many areas, being always immediately available, being able to deliver information in a variety of forms (spoken, written, graphical, animated, etc.), and being tireless, undistractable, and invariably in helpful mode.

A distinction that is likely to prove to be important in the future is that between search engines designed for general use and those that are customized to specific users. This is a special case of the more general distinction between generic and individualized tools. Currently available search engines are examples of the generic type. They serve one user as well as another. We can imagine, however, information-finding programs with the capability to learn to function as agents, or alter-egos, for individual users. Such an agent would be able to find its way around the various information resources that exist and, with a detailed model of its owner's knowledge and interests in hand, identify items its owner would want to know about, without always having to be told.

A foreshadowing of this idea can be found in the writings of such visionaries as Vannevar Bush and J. C. R. Licklider, and a more recent expression of it is encapsulated in the notion of a knowledge robot, or *Knowbot,* a term coined and registered as a trademark by the Corporation for National Research Initiatives. Knowbots are "programs that move from machine to machine, possibly cloning themselves.... They communicate with one another, with various servers in the network and with users. In the future, much computer communication could consist of the interactions of Knowbots dispatched to do our bidding in a global landscape of networked computing and information resources" (Cerf, 1991, p. 74).

Imagine having an agent to which one could assign tasks much as one would to a human assistant, but that had immeasurably greater capacity, energy, access to information resources, and patience. With the current state of the art, it should be possible to have an agent that a researcher might ask, for example, to find the 10 most frequently cited publications on a specified topic that were published in the past 10 years, or to list all the articles that reference a specified publication, and to retrieve the abstracts or full articles of the titles one finds interesting. Much more sophisticated searches could be done with agents that had enough intelligence to read text with a modicum of understanding.

Real-Time Tutorial Help

We may distinguish two types of tutorial help. One is the immediate help one can get for understanding something one is reading via online look up of definitions, synonyms, explanations, elaborations, and so on. The other is the kind of tutoring one might expect from a human tutor who is charged with the task of helping one become an expert in a given area. The first type is the more easily implemented of the two and already exists in rudimentary form in some systems. The second type is not more than a vision at the present, but it is one that seems plausible in the foreseeable future.

One of my fantasies for the future is that of a personalized computer system that can serve the roles of memory augmenter, fact finder, tutor, and alter ego. I have in mind a system that has expert knowledge in many areas, speech capability (understanding and production), the ability to generate and refine a model of its user, and very high-level tutorial capability. Suppose that such a system were available and one wished to use it to learn about, say, Italian Baroque music. The computer in my fantasy not only would have encyclopedic knowledge of the subject, but would know—having discovered—precisely what its user knew of the subject as well. It would know or be able to determine, for example, that its user knows something about Nardini and Tartini, but nothing about Pergolesi or Albinoni. It would be able not only to answer questions but to volunteer information, and the information it volunteered would be appropriate to its user's current level of understanding. It could supplement what it has to say with illustrations of the work of specific composers—with the sounds that would have been produced by instruments of the day. By constantly checking its model of its user's knowledge, it could offer new information to fill in existing gaps, sharpen fuzzy distinctions, generate new insights, and so on. This fantasy is not likely to be a reality soon, but neither is it so far removed from what is currently possible to be considered intrinsically impossible in the foreseeable future.

Memory Aids

People need memory aids for various purposes. In some cases, what is needed are ways of extending or supplementing human memory—to reduce the need for relying on human memory or to provide a storage medium of greater reliability and capacity, or to organize information to maximize the ease with which it can be retrieved for future use. Sometimes what is needed is a system that can help compensate for memory loss or dysfunction as a result of head injuries, strokes, and other types of traumas, as well as of normal aging.

Often what one needs by way of a memory aid is a means for capturing information or ideas for future reference. Some people (e.g., researchers, intelligence analysts, journalists) are likely to find it necessary to read extensively, searching for information that is relevant to specific problems on which they are working. Often they come across information that, although not directly useful for their immediate problem, is of general interest and likely to be useful at some future date. For a long time the standard way of capturing such information has been with the use of notebook and pencil. But notes on paper have a way of getting lost, and even when they do not, they are often difficult to organize in such a way as to facilitate later retrieval when they are likely to be of use.

An alternative to the paper notebook, and one that is being used increasingly as more and more people become computer users, is that of maintaining computer-based *incubation* or *fragment* files, in which interesting and potentially useful information is stored when it is discovered. Such files not only serve as convenient repositories of the information but facilitate organization, reorganization, and selective retrieval.

Numerous efforts have been made to develop techniques that are effective in restoring, at least partially, lost or diminished memory function. Many of these techniques make use of technology, and especially, computer technology (Glisky, 1995; Kerner & Acker, 1985; Parenté & Anderson, 1991; Walker, 2001).

Prospective Memory Aids (Reminders)

Most of us use a variety of techniques to aid our prospective memories— our intentions to do certain things at specific times. We write notes to ourselves, keep to-do lists, mark calendars, ask a spouse or friend to remind us, etc. One of the roles that a memory amplification system might perform is that of a memory lackey. Ideally, one would like to have a system to which one could say "Remind me to . . . ," and then have the system, on its own initiative, do the reminding at the appropriate time. The problem with the notebook system that many of us use is that one must remember to look at it, and to do so at short-enough intervals to ensure one does not see a reminder only after its usefulness is past.

It is easy to imagine various ways in which computer technology could be used to serve the reminder function. Hand-held and pocket-sized devices are of special interest in this regard (Herrmann, Sheets, Wells, & Yoder, 1997; Herrmann, Yoder, Wells, & Raybeck, 1996). If one habitually reads e-mail daily, one should be able to send oneself messages with to-be-delivered dates so they would appear in one's mailbox on the days one wishes to receive specific reminders. A more ambitious system might involve the use of a paging beeper, or a vibrotactile stimulator (perhaps attached to a wristwatch band) that could be actuated by a radio signal to let one know that a reminder has just been delivered to one's e-mail box. It is essential that such aids be active and not depend on the user taking the initiative to check them periodically.

Inferencing Aids

The psychological literature is replete with studies the results of which have been taken as evidence of numerous ways in which human reasoning goes astray. Examples of fallacies, biases, and other predilections to irrationality that have been deemed to characterize human reasoning include the gambler's fallacy, the confirmation bias, the attribution bias, and hasty generalizations. One could generate a long list, and several writers have done so

(Gilovich, 1991; Kahneman, Slovic, & Tversky, 1982; Nisbett & Ross, 1980; Piattelli-Palmarini, 1994).

Whether the reasoning that people typically do is as badly flawed as some investigators have claimed is debatable. However, there can be little doubt that there is room for improvement. The question of interest for present purposes is what kinds of assistance might have material beneficial effect. Possibilities include: prompting the articulation of unstated assumptions, identifying ambiguities, noting counterfactual possibilities, identifying potentially falsifying evidence—to name a few. One can imagine a system that would help one formulate an argument: by asking questions that demand the explication of unstated assumptions, the specification of alternative assumptions that might be made, the expression of evidential reasons for assertions, the consideration of unstated implications of conclusions, and so on.

The purpose would not necessarily be to change the user's beliefs, but rather to help him or her to get a better understanding of what particular beliefs entail, which includes an appreciation of unstated assumptions on which they may rest, of evidence that would disconfirm them, and of arguments that might be made against them. People may be surprised when they discover unstated assumptions that underlie a belief, or what a belief may imply about other beliefs one must hold in order to be consistent.

People may find it difficult to state reasons for beliefs either because they lack them or because they have had them but cannot recall them. And sometimes they may have reasons they are unwilling or reluctant to admit, perhaps even to themselves. The tendency to interpret data in ways that are consistent with favored hypotheses and to fail to properly weigh, or even to consider, alternative interpretations appears to be strong and quite pervasive (Nickerson, 1998). Having to explicate reasons for beliefs should improve the quality of reasoning in several respects.

Techniques for conveying the notion of weight of evidence, in contrast to binary—yes/no, either/or—thinking should be helpful. Sometimes binary choices are forced, but until they are, one should continue to weigh evidence and remain open to the possibility of shifts, and especially to the possibility that the favored hypothesis could turn out to be wrong. When a binary choice is forced, it can be made on the basis of the balance of evidence at the time.

Communication

Personal communication between and among people has already been affected greatly by the rapidly growing use of electronic mail, electronic bulletin boards, electronic chat rooms, and other computer-mediated communication innovations (Kiesler, 1997; Sproull & Kiesler, 1991; Wood

& Smith, 2001). The same technology has also been changing the way scientific information is disseminated (Stix, 1995). More and more, the results of scientific research are being distributed electronically almost as soon as they have been obtained. Some of this distribution is informal, involving colleagues who, in the past, habitually exchanged preprints of work in progress before submitting for formal publication in refereed journals. However, increasingly scientific journals and newsletters are making use of network technology to provide access either to refereed articles or to supplementary information. Many journals that have not yet gone electronic completely are accepting (or requiring) electronic submission of manuscripts and handling the entire review process electronically.

Collaboration and Corporate Decision Making

Group decision making can be influenced by uncertainties about relevant facts, as well as by emotions, and differences in status and persuasive abilities of the participants. Ensuring that decisions are based firmly on the merits of a situation requires minimizing the effects of such factors.

Imagine a system designed to help a group converge on a consensus. Assume that the objective is to get a decision that makes optimal use of the aggregate knowledge and expertise of the group as a whole. This probably means that all participants in the decision making need to know what the relevant facts are and what everyone else really thinks. The system then should ensure: that all relevant knowledge be considered; that all opinions get expressed without fear of ridicule or reprisal; that the data gathering, opinion-shaping process stay open as long as necessary; and that individuals do not publicly commit prematurely to a position and then feel compelled to defend that position.

In order to realize these objectives, the system should maintain the anonymity of the sources of ideas and arguments. (At least it should have that option. A question that deserves empirical study is that of how the process, and its outcomes, might differ depending on this factor.) It should preclude domination by forceful personalities and intimidation because of position within an organizational hierarchy. It should provide for straw votes on issues with secret ballots. Maintaining anonymity within a group of colleagues or acquaintances is probably not trivially easy even when the communication is through computer terminals and not face-to-face, but it is probably doable. Identities might be masked and emotional language neutralized by the use of filters—voice digitization, paraphrasing, etc. The system should tap information resources to answer questions of fact, to confirm or correct factual claims, to resolve factual disputes (or admit to being unable to do so). It should have the ability to judge the degree of consensus within the group on a specified issue, based on polling data, and perhaps to suggest compromises or strategies that would increase the consensus.

The imaginary system just described is for a truly democratic decision-making process. A democratic process may be desirable in some situations and undesirable in others. The owner of a privately held company, for example, may understandably be unwilling to have all the decisions for his or her company made by committee. Nevertheless, such an owner might well like to have the benefit of the output from such a process to inform the decisions he or she makes.

To my knowledge, such a system does not exist, but creation of something like it is not beyond current technology. It is one example of many that could be given of what is sometimes referred to as *groupware* or "software that helps groups communicate, cooperate, coordinate and solve problems" (Miller & Yesford, 2001). Computer-supported cooperative work by groups or teams has been a focus of attention among researchers in human factors and human–computer interaction for some time, as evidenced by special issues on the topic by several journals (e.g., *Human-Computer Interaction*, 1992, *Interacting With Computers*, 1992, and *International Journal of Human-Computer Interaction*, 2002). Much of the collaborative work that has been done with the help of such systems has been in scientific fields, but the technology is being used in the humanities as well (Inman, Reed, & Sands, 2003).

Other

There are many ways in which information technology is being, or could be, used to amplify cognition. Space constraints preclude discussing more than a few of them. Here I will simply list, with very brief comment, what seem to me to be the more apparent examples beyond those already mentioned.

- Facilitation of idea generation. Computer-assisted brainstorming in which ideas are collected anonymously and without assessment until the generation phase is completed.
- Knowledge assessment. Knowledge-probing techniques to determine by adaptive sampling what anyone (including oneself) knows on a given subject.
- Debiasing. Antidotes to common biases and illogicalities (confirmation bias, hindsight bias, attribution bias, gambler's fallacy, egocentric biases).
- Value discovery. Often we do not know our own values and their relative weights as they pertain to choice situations, but they can be inferred from expressed preferences between hypothetical alternatives.

- Decision problem elucidation. Facilitation of identification of possible states of the world and of decision alternatives.
- Problem-solving help. Prompting, and illustrating, the use of heuristics (extreme cases, decomposition, working backwards), suggesting analogies, helping find useful representations, keeping track of partial solutions.
- Prediction. Provision of runnable "what if" models that let a decision maker explore possible (probable) consequences of possible actions.
- Error prevention. Automatic checking of an electronic *Physician's Desk Reference*, for example, for advisability of a particular drug prescription, taking into account possible interactions with other drugs being taken.
- Negotiation and conflict resolution. Removing (or at least masking) emotion, through an impersonal bid–counterbid process, in attempting convergence on mutually acceptable terms for agreement.
- Facilitating probabilistic reasoning. By explicating the differences between joint and conditional probabilities, providing actuarial data to help make realistic probability estimates, etc.
- Planning, forecasting, budgeting. Forcing identification and consideration of possible contingencies as a means of addressing the ubiquitous problem of underestimation of time and costs of planned projects.

RISKS IN COGNITION AMPLIFICATION TOOLS

There are risks involved in the development of cognition amplification tools. There is, for example, the possibility of cognitive technology being used in exploitive ways—for propaganda, brainwashing, and manipulation. Use of the Internet to facilitate identity theft, distribution of child pornography, false advertising, and other forms of cyber crime is already a reality and appears to be on the increase. Tools that greatly increase the accessibility of information also sometimes facilitate the invasion of privacy. But such is the price of progress; any tool can be used for bad purposes as well as good, and the more powerful the tool, the greater the potential in both cases.

It is reasonably clear that information technology already provides many people with easy access to resources that can be very helpful to their efforts to prosper in a competitive world. And the resources that can be tapped—by some of those who can benefit from them—will increase. But many people who could benefit from such resources do not have access to them. This is especially true in third-world countries, where illiteracy and economics conspire to ensure that many of the neediest people have little, if any, access to the resources that information technology represents. Many observers have noted this problem and the likelihood that the disparity between the haves and the have-nots in this regard will increase. There have been attempts to address the problem and devices that provide limited access at relatively low cost have been produced (Harvey, 2002), but much more needs to be done if the people who have least access to the technology are not to fall farther and farther behind.

E-mail has greatly facilitated communication among people around the globe; it has also made users vulnerable to spam (junk e-mail) and to computer viruses, worms, and other forms of electronic mayhem that malicious hackers are able to conceive. The ability to store very large quantities of information—correspondence, financial information and other personal records, manuscripts (by writers), blueprints (by architects and builders), data (by experimenters), and so on—is a great convenience, but information so stored is very easily lost or destroyed. The availability of versatile data-analysis software that can be used with a modicum of understanding of the rationales on which the offered analyses rest makes inappropriate and inept use as easy as justified use.

Word-processing software has unquestionably facilitated the processes of composition and editing enormously, but has it improved the quality of the written material that is produced? Some might argue that at least in some instances it has done precisely the opposite. More generally, there is a question of the extent to which applications of technology to the simplification of tasks may inhibit the development of skills that would be necessary if the technological "crutch" were not available. I recently made a small-item purchase at a hardware store. My bill came to a total of $5.27. I gave the clerk $10.00. After the clerk rang up the purchase, I said I would give him the 27 cents, thinking that would simplify things for him and relieve me of accumulating in my pocket more small change. The clerk said it was too late, the machine had already recorded the sale and showed the amount of change I had coming. I saw no point in arguing. I do not mean to suggest that simplifying cognitive tasks through the application of technology is always a bad idea—clearly it is not—but we should be alert to the fact that when the technology makes specific skills obsolete in practical situations, it undoubtedly decreases the likelihood that those skills will be acquired and retained. That may be acceptable in the case of some skills and not in that of others.

There is always a cost/benefit ratio question. E-mail is a great convenience, but if one is not careful, it can also become a burden. Having immediate access to everyone in the world is a benefit; everyone in the world having immediate access to you is less clearly so. Trying to protect oneself from spam, viruses, and unwanted demands on one's time can be time consuming. Keeping information current in a personal data manager can itself become an information management problem. And so on. Whether the benefits outweigh the costs in particular instances must be decided on a person-by-person basis. In my own case, I am reasonably certain they do, but I am sensitive to the need for some vigilance to ensure that it remains so.

CONCLUDING COMMENT

Was there ever a time when human beings existed and they did not use technology, broadly defined, to aid cognition? Presumably as soon as humans learned to count and to measure, they made devices to help them do so and to remember the results. The development of symbol systems and written language was certainly among the most noteworthy technological achievements of prehistory; there is no other technological advance whose effects on human history rival those of this one. But there are countless examples of artifacts (devices, systems, procedures) that have been invented throughout history to facilitate the performance of cognitive tasks or to amplify human cognition in one or another way.

In this chapter, I have tried to illustrate the use of artifacts to aid cognition by focusing on sliding rules and other similar devices that have been developed to aid the performance of a variety of computing tasks. Bringing this story up to date leads us, of course, to the modern digital computer. *Computer*, however, is a misnomer for the devices that we refer to by that name, because computing, in the usual sense of that word, does not begin to capture what this device can be made to do. In the latter part of this chapter, I have tried to identify current needs and opportunities for cognitive amplification and to note some ways in which information technology has been, or could be, used to respond to them.

The relationship between technology and cognition is one of dependency that goes both ways. There would be little in the way of technology in the absence of cognition. And cognition would be greatly handicapped if all its technological aids were suddenly to disappear. Technology is a product of cognition, and its production is a cyclic, self-perpetuating process. Cognition invents technology, the technology invented amplifies the ability of cognition to invent additional technology that amplifies further the ability of cognition . . . and so it goes.

REFERENCES

Anonymous (1880). *The carpenter's slide rule: Its history and use.* Birmingham, UK: John Rabone & Sons. (Reprinted by Ken Roberts, Fitzwilliam, NH)

Ashby, W. R. (1956). Design for an intelligence amplified. In C. E. Shannon & J. McCarthy (Eds.), *Automatic studies* (pp. 215–234). Princeton, NJ: Princeton University Press.

Bell, E. T. (1937). *Men of mathematics.* New York: Dover.

Bush, V. (1945, July). As we think. *The Atlantic Monthly*, pp. 101–108.

Cajori, F. (1994). *A history of the logarithmic slide rule and allied instruments.* Mendham, NJ: Astragal Press. (Original work published in 1910)

Cerf, V. G. (1991). Networks. *Scientific American, 265*(3), 72–81.

Engelbart, D. C. (1963). A conceptual framework for the augmentation of man's intellect. In P. W. Howerton & D. C. Weeks (Eds.), *Vistas in information handling* (Vol. 1, pp. 1–29). Washington, DC: Spartan Books.

Gilovich, T. (1991). *How we know what isn't so: The fallibility of human reason in everyday life.* New York: The Free Press.

Glisky, E. L. (1995). Computers in memory rehabilitation. In A. D. Baddeley, B. A. Wilson, & F. N. Watts (Eds.), *Handbook of memory disorders* (pp. 557–575). Chichester, UK: Wiley.

Harvey, F. (2002). Computers for the third world. *Scientific American, 287*(4), 100–101.

Herrmann, D. J., Sheets, V., Wells, J., & Yoder, C. Y. (1997). Palmtop computerized reminding devices: The effectiveness of the temporal properties of warning signals. *AI and Society, 11*, 71–84.

Herrmann, D. J., Yoder, C. Y., Wells, J., & Raybeck, D. (1996). Portable electronic scheduling/reminding devices. *Cognitive Technology, 1*(1), 36–44.

Hopp, P. M. (1999). *Slide rules: Their history, models, and makers.* Mendham, NJ: Astragal Press.

Ifrah, G. (1987). *From one to zero: A universal history of numbers* (L. Blair, Trans.). New York: Viking/Penguin.

Inman, A., Reed, C., & Sands, P. (Eds.). (2003). *Electronic collaboration in the humanities: Issues and options.* Mahwah, NJ: Lawrence Erlbaum Associates.

Jourdain, P. E. B. (1956). The nature of mathematics. In J. R. Newman (Ed.), *The world of mathematics, Vol. 1* (pp. 4–72). New York: Simon & Schuster. (Original work published in 1913)

Kahneman, D., Slovic, P., & Tversky, A. (Eds.). (1982). *Judgment under uncertainty: Heuristics and biases.* Cambridge, UK: Cambridge University Press.

Kerner, M., & Acker, M. (1985). Computer delivery of memory retraining with head injury patients. *Cognitive Rehabilitation, 6*(26), 26–31.

Kiesler, S. (Ed.). (1997). *Culture of the Internet.* Mahwah, NJ: Lawrence Erlbaum Associates.

Licklider, J. C. R. (1960). Man–computer symbiosis. *Institute of Radio Engineers Transactions on Human Factors Electronics*, HFE-1, 4–11.

Lusted, H. S., & Knapp, R. B. (1996). Controlling computers with neural signals. *Scientific American, 275*(4), 82–87.

Maor, E. (1994). *e: The story of a number.* Princeton, NJ: Princeton University Press.

Menninger, K. (1969). *Number words and number symbols: A cultural history of numbers.* Cambridge, MA: MIT Press.

Miller, L. A., & Yesford, D. (2001). The Innovator™: A system for improving group effectiveness. *Cognitive Technology, 6*(1), 41–44.

Nickerson, R. S. (1986). *Using computers: Human factors in information technology.* Cambridge, MA: MIT Press.

Nickerson, R. S. (1988). Counting, computing, and the representation of numbers. *Human Factors, 30*, 181–199.

Nickerson, R. S. (1994). Electronic bulletin boards: A case study of computer-mediated communication. *Interacting With Computers, 6*, 117–134.

Nickerson, R. S. (1995). Human interaction with computers and robots. *The International Journal of Human Factors in Manufacturing, 5*, 5–27.

Nickerson, R. S. (1998). Confirmation bias: A ubiquitous phenomenon in many guises. *Review of General Psychology, 2*, 175–220.

Nicolelis, M. A., & Chapin, J. K. (2002). Controlling robots with the mind. *Scientific American, 287*(4), 46–53.

Nisbett, R. E., & Ross, L. (1980). *Human inference: Strategies and shortcomings of social judgment.* Englewood Cliffs, NJ: Prentice-Hall.

Pachham, J. (1997). Barrel gauging. *The Chronicle of the Early American Industries Association, 50,* 121–124.

Parenté, R., & Anderson, J. (1991). *Retraining memory: Techniques and applications.* Houston, TX: CSY.

Piattelli-Palmarini, M. (1994). *Inevitable illusions: How mistakes of reason rule our minds.* New York: Wiley.

Roberts, K. D. (1982). Introduction. In *Reprint of the carpenter's slide rule: Its history and use.* Fitzwilliam, NH: Ken Roberts.

Ramo, S. (1961). The scientific extension of the human intellect. *Computers and automation, 10,* 9–12.

Routledge, J. (1867). *Instructions for the use of the practical engineers' and mechanics' improved slide rule.* Birmingham, UK: John Rabone & Sons.

Salomon, G. (Ed.). (1993). *Distributed cognitions: Psychological and educational considerations.* New York: Cambridge University Press.

Sass, M. A., & Wilkinson, W. D. (Eds.). (1965). *Computer augmentation of human reasoning.* Washington, DC: Spartan Books.

Schmandt-Besserat, D. (1978). The earliest precursor of writing. *Scientific American, 238*(6), 50–59.

Sproull, L., & Kiesler, S. (1991). *Connections: New ways of working in the networked organization.* Cambridge, MA: MIT Press.

Stix, G. (1995). Toward "Point One." *Scientific American, 272*(2), 90–95.

Walker, W. R. (2001). External memory aids and the use of personal data assistants in improving everyday memory. *Cognitive Technology, 6*(2), 15–25.

White, L., Jr. (1962). The act of invention: Causes, contexts, continuities, and consequences. *Technology and Culture, 3,* 486–500.

Wood, A. F., & Smith, M. J. (2001). *Online communication.* Mahwah, NJ: Lawrence Erlbaum Associates.

2

Technology and the Development of Intelligence: From the Loom to the Computer

Ashley E. Maynard
University of Hawaii

Kaveri Subrahmanyam
California State University, Los Angeles

Patricia M. Greenfield
UCLA

The nature of a culture's tools at a particular time influences that culture's operational definition of intelligence. That is, the cognitive skills required to develop and utilize a culture's tool set become an important component of a group's implicit definition of intelligence. The major thesis of this chapter is that using a particular tool set develops the cognitive skills that are part of a group's implicit definition of intelligence. Just as we embed cognitive skills that are important in utilizing our own culture's tools in our own intelligence tests, so too we can imagine that the intelligence tests of other cultures might reflect their own cultural tools (Greenfield, 1998). This chapter will show that different tools in different cultures not only utilize, but also develop, particular sets of cognitive skills. Tools themselves evolve through historical time and thus reflect the social and cognitive developments at a particular point in history in a particular place, and at the same time they influence these developments. Therefore, when cultural tools change, it follows that cognitive skills and valued forms of intelligence should change as well; and such cultural change will be one focus of this chapter.

Humans, and primates more generally, are considered a tool-using species. The tool-making and tool-using capacities of humans are skills/abilities that have developed over the course of human evolution and continue to evolve (Boyd & Silk, 2000). If we define intelligence as successful behavioral adaptation to an ecological niche (see Scheibel, 1996), tools are an essential component of the human adaptation to a variety of different niches. Just as tools adapt to ecological niches, cognitive skills adapt to tools and to the social practices in which they are embedded (Saxe, 1994).

Let us stop a moment to define some terms. How do tools relate, definitionally, to technologies, the topic of this book? Tools are the byproduct of technologies, that is, of an underlying knowledge base. For example, the technology of literacy creates the book as a specific tool. The technology of electronics creates the computer as a specific tool. In this chapter, our particular focus is on the ways in which a culture's technologies, and the tools that are components of these technologies, both influence and reflect the forms of intelligence that are developed and valued in that culture.

Bruner (1966), along with Cole and Griffin (1980), believes that the development of intelligence is to a great extent the internalization of the tools of the cultural niche in which the child or person operates. Vygotsky (1962, 1978) makes a comparison between symbolic tools and physical tools that highlights the theoretical position of the tool in cognition. Lave (1977), Nunes, Schliemann, and Carraher (1993), Saxe (1999), and Guberman (1996) emphasize the role of tool-based activities in the development of cognitive representations and operations. This perspective on tools is based on a fundamental idea in Vygotskian psychology, well expressed by the Russian psychologist O. K. Tikhomirov: "Tools are not just added to human activity; they transform it" (1974, p. 374). In this view, tools can be either concrete or symbolic. Most tools are a mixture of the two, in that even a seemingly concrete tool like a loom requires symbolic operations.

Technology develops intelligence through the internalization of cognitive skills required by various tool systems. In some cases, physical skills carried out with the aid of a tool become mental skills. Often seemingly physical skills require particular cognitive operations of varying levels of complexity. In other cases, mental skills carried out with the aid of a tool can later be performed independently (Salomon, 1988). In other words, technology operates in what Vygotsky (1962, 1978) called the "zone of proximal development": the area between aided and independent cognitive achievement. In summary, technologies develop and lead to the internalization of the mental skills that they require for their utilization, and these skills are then embedded in a culture's implicit definition of intelligence.

At this point we must digress to mention that there are, in fact, two major categories of intelligence that exist around the world: social intelligence and technological intelligence (Mundy-Castle, 1974). Tool systems have their primary impact on one of these: technological intelligence, or cognitive skills relating to the world of things. We must recognize that although both exist in every society, they receive differential emphasis; some cultures emphasize social intelligence more, whereas others put greater emphasis on technological intelligence (Greenfield, Keller, Fuligni, & Maynard, 2003; Mundy-Castle, 1974). Our point is that the skills required for a culture's tool systems become that culture's implicit definition of technological intelligence, whether or not technological intelligence is the most important type of intelligence in a particular culture. In the Vygotskian approach, technological intelligence is implicitly assumed to be the only category of intelligence. With the aid of Mundy-Castle (1974), Wober (1974), Sternberg, Conway, Ketron, & Bernstein (1981), and Dasen (1984), we have come to realize that this is not the case. We will return to this point later.

Learning to use a particular technology both utilizes and develops mental skills on different levels of cognition. The three levels on which we focus in this chapter are attention, representation, and mental transformation (based on Piaget's notion of concrete operational intelligence). These levels go from the lower, more automatic levels to the higher, more intentional levels of cognition. We explore how these three cognitive levels of intelligence are influenced and used by different types of technology. Based on our own empirical research programs, we have chosen to focus on two types of technology in two different parts of the world: computer technology in the United States and weaving in Maya Mexico. Whereas computer technology is an invention of the 20th century, the backstrap loom used by the Maya has a history that goes back more than 4,000 years in the Americas (Greenfield, 2004). Both weaving and the computer constitute examples of technologies that not only reflect but also develop a culture's valued forms of intelligence. Each profile of highly developed cognitive skills illustrates the close connection between a tool system and the development of technological intelligence on all three levels of cognition.

COMPUTER TECHNOLOGY AND COGNITIVE SKILLS

Media are symbolic tools that vary from culture to culture and from one historical period to the next. Olson and Bruner write that "each form of experience, including the various symbolic systems tied to the media, produces a unique pattern of skills for dealing with or thinking about the world. It is the skills in these systems that we call intelligence" (1974, p. 149).

One symbolic tool that began to change the landscape of home, education, and workplace in the United States and other parts of the world at the end of the 20th century is the interactive technology of the computer. How has computer technology affected the cognitive skills that we call intelligence? In this section, we consider two of the most popular of the computer applications, games and the Internet, in order to address this question.

Greenfield argued that computer applications such as action games require and develop a different profile of cognitive processes compared to earlier modes of communication, such as print (e.g., Greenfield, 1984a, 1985). Indeed there seem to be a whole set of literacy skills associated with computers and the video screen that are quite distinct from the traditional literacy skills required for print (see, e.g., Greenfield, 1984a, 1987, 1990a, 1998). Most computer applications have design features that shift the balance of required information processing from verbal to visual (Subrahmanyam, Greenfield, Kraut, & Gross, 2001). For instance, action video games, which are spatial, iconic, and dynamic, have multiple, often simultaneous, things happening at different locations and the ability to "read" and utilize the information on computer screens may therefore require a variety of attentional, spatial, and iconic skills.

The suite of skills children develop by playing such games can provide them with the training wheels for computer literacy and can help prepare them for science and technology, where more and more activity depends on manipulating images on a screen. Research has provided evidence for the thesis that computer game playing can have an impact on specific cognitive skills. Although the term *cognitive skills* encompasses a broad array of competencies, most of this research has focused on components of visual intelligence, such as perception and attention, representation (iconic and spatial), and mental transformations. These skills are crucial to most video and computer games, as well as to the Internet and many other computer applications (Greenfield, 1984a).

Attention

We begin our survey with the attentional level of cognition. On this level, one important skill involved in playing computer and video games is divided visual attention, sometimes called parallel visual processing. This is the skill of keeping track of multiple things happening simultaneously. In almost all action games, more than one entity is present and acting on the screen at the same time. This characteristic goes back at least 2 decades to the maze game of Pac-Man, one of the first popular action video games. Skilled play at Pac-Man requires simultaneously keeping track of the Pac-Man character, four monsters, your location in the maze, and four energizers. Many other more complex games, past and present, have even more information sources that must be dealt with simultaneously (Greenfield, 1984a). In order to be a successful player, one must monitor more than one location on the screen. Would this technology-based game requirement translate into skill in parallel visual processing? Would practice in the software tool of action games produce skill in dividing visual attention?

Greenfield, deWinstanley, Kilpatrick, & Kaye (1994) explored the effect of video game expertise and experience on strategies for dividing visual attention among college students enrolled in introductory psychology. Divided attention was assessed by measuring participants' response time to two

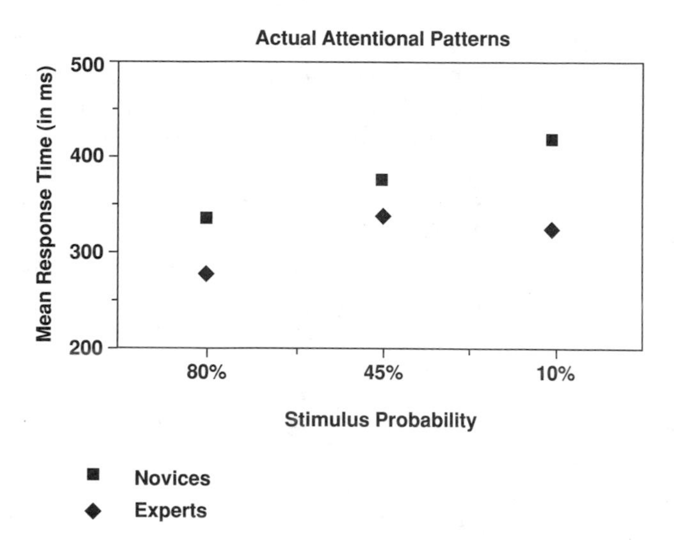

Fig 2.1. Actual relationship between video game expertise and strategies for dividing attention (Greenfield, deWinstanley, Kilpatrick, & Kaye, 1994).

events of varying probabilities at two locations on a computer screen. In one condition, a target appeared more often at one location than another. In another condition, a target appeared with equal probability at both locations. For each condition, participants were told in advance what the probabilities were. Translating these known probabilities into monitoring strategies is analogous to what one must do after inducing the differential probabilities of various events at various locations in an action video or game.

Participants who were expert game players (those who scored higher than 200,000 on the game Robot Battle) had faster response times than novices (those who scored below 20,000 on the game Robot Battle) at both high- and low-probability positions of the icon (see Fig. 2.1). (Note that these groups represent only the extremes of game expertise.) The strategic nature of the expert players' attentional skill (versus a simple improvement in eye–hand coordination) was shown in the patterning of their performance. Prior research showed that, in comparison to equiprobable targets, people generally allocate more attention (and respond faster) to a high-probability target, whereas they allocate less attention (and respond slower) to a low-probability target (Posner, Snyder, & Davidson, 1980). Correlatively, expert players were faster only in response to the high- and low-probability targets; there was no difference between the groups in the equiprobable condition, which presumably does not require strategic deployment of attentional resources. Second, relative to the condition with two equally probable targets, expert players showed no decrement in skill in the low-probability condition; novice players, in contrast, did.

Even more important, the researchers were able to establish a causal relationship between playing an action game and improving strategies for monitoring events at multiple locations. In a second experiment, introductory psychology students (unselected for video game skill) were randomly assigned to play the action arcade game Robotron or to be in a no-play control group. Robotron, like Robot Battle, involves multiple entities acting simultaneously. The attentional task remained the same and was administered as both pretest and posttest. In the pretest, more- and less-experienced players (not as extremely different as the novices and experts in the first study) differed only at the high-probability target, where the more-experienced players again had significantly faster response times; there was no difference at the low-probability target. After 5 hours of playing Robotron, members of the experimental group responded significantly faster to the target at the low-probability position on the screen; in contrast, members of the control group, who also took the attentional posttest, did not show this improvement. Practice on the test by itself (the control condition) yielded selective improvement with the equiprobable targets, which require less strategic skill and, in the first study, registered no difference between expert and novice players. Overall, the studies showed (a) that experts at utilizing the game technology had better developed strategies for dividing visual attention than did novices and (b) that practice with this technology improves strategic competence in monitoring events at a relatively improbable location.

Recent research confirms Greenfield et al.'s finding regarding the effect of video game playing on attentional skill. Green and Bavelier (2003) reported that adult video game players (who had played action video games on at least 4 days per week for a minimum of 1 hour per day for the previous 6 months) had enhanced attentional capacity compared to nonvideo game players (who had little or no video game usage in the past 6 months). The attentional skills were assessed using an enumeration task (reporting the number of squares in a briefly flashed display), a flanker compatibility effect (the effect of a distractor on a target task), and a modification of the "useful field of view" task (measures the ability to locate a target among distractors to assess attention over space).

In addition, Green and Bavelier provided action video game training to a group of nonvideo game players by asking them to play the action game Medal of Honor for 1 hour per day for 10 consecutive days; a control group was asked to play the game Tetris for the same time span. Tetris is a dynamic puzzle game in which only one event takes place at a time; in contrast, Medal of Honor is a battle game in which multiple entities are simultaneously engaged in various actions. The results suggested that Medal of Honor led to greater improvements in attentional strategies on all the tests than did Tetris.

The transfer effect of video game playing obtained by Green and Bavelier on entirely different attentional tasks is noteworthy. Also noteworthy are the very consistent effects in both the correlational and experimental study and the effects across a wide range of attentional tasks. Although the two pairs of studies cannot be directly compared, we think the greater consistency of effects in Green and Bavelier's studies is due in large part to the fact that the study was carried out a good decade later. In the intervening time home video sets, computers, games for younger children, and hand-held games had become pervasive in U.S. homes, allowing participants more prior experience with electronic games, and, most important, more experience earlier in their development (Vanderwater, Wartella, & Rideout, 2003). We believe that earlier practice with a technology will lead to a larger impact on the cognitive skills that define intelligence in a particular culture, as well as to more precocious development of those skills (LeVine, 2002).

Finally, strategies for dividing visual attention have come to be necessary for handling recently developed computer formats now omnipresent on television, as well as on the Internet. On TV, there are divided screens with textual information running across the bottom and, on financial programs, down one side of the screen, all while a talking head holds forth in the rest of the screen space. On the Internet, teenagers (and others) often move from window to window, simultaneously monitoring their instant messages, e-mail, and homework, all while downloading music videos (Gross, 2003). Skills developed in video games can be useful in monitoring these contexts that also require parallel processing.

Representation

Experience with computer video games has also been found to affect the development of mental representation skills. Salomon (1988) asserted that symbolic forms in computer tools can be internalized as cognitive modes of representation as a person interacts with a computer. Both iconic and spatial representation are crucial to scientific and technical thinking; these modes of representation enter into the utilization of all kinds of computer applications.

Iconic Representation. One important representational skill embodied in computer games is iconic or analog representation—or the ability to create and read images such as pictures and diagrams. Indeed iconic images are frequently more important than words in many computer games. Greenfield, Camaioni, et al. (1994) found that playing a computer game shifted representational styles from verbal to iconic. In the study, undergraduate students played the game Concentration either on a computer or on a board.

The goal of the game was to open either virtual or real doors to identify the location of pairs of numerals. The computer version was comprised of icons: virtual doors and a cursor in the shape of a hand. The board version had no icons, but involved direct action on an object—the participant used his or her hand to lift a solid door in order to reveal a numeral. A pretest and posttest included several dynamic video displays from Rocky's Boots, an educational computer simulation designed to teach the logic of computer circuitry; Fig. 2.2 provides an example of what the participants saw. Their task was to try to figure out what was going on; they were given no clues as to the content or operation of the displays.

When asked to explain on a pencil-and-paper test the operation of displays such as the one in Fig. 2.2, those who had played the game on the computer offered more iconic diagrams in their descriptions, whereas those who played the game on a board offered more verbal descriptions (see Fig. 2.3). Thus, playing a computer game that used icons influenced participants to use icons in their representations; the game technology had shifted the construction of representations from verbal to iconic.

This study was a cross-cultural one, comparing students in Rome, Italy, where computer technology, at that time, was much less diffused, to students in Los Angeles, where the technology was much more diffused. Participants in Los Angeles preferred to use diagrams or icons compared to the Italians, who used words in responding to the test (see bottom of Fig. 2.4). We see this as a correlational finding that indicates the ecological validity and generality of the experimental result. In other words, technology appears to operate in the real world, not just in a specific experimental setting.

But not only did the technology make participants create more iconic representations, it also seemed to make participants understand iconic representation better. Corresponding to their relative exposure to video games, experienced players, Americans, and males understood the dynamic iconic simulations of the logic of computer circuitry presented on a video screen better than did inexperienced video game players, Italians, or females (see top of Fig. 2.4). These correlational data indicate that computer technology, as instantiated in action games, not only increases the frequency of iconic representation, it also increases comprehension of this mode of representation.

Spatial Representation. Spatial representation is considered a domain of skills rather than a single ability (Pellegrino & Kail, 1982) and includes skills such as mental rotation, spatial visualization, and the ability to deal with two-dimensional images of a hypothetical two- or three-dimensional space. Spatial representational skills are used in all kinds of computer applications, including word processing, programming, and the recreational medium of action video games (Gomez, Bowers, & Egan, 1982; Greenfield, 1983, 1984a,

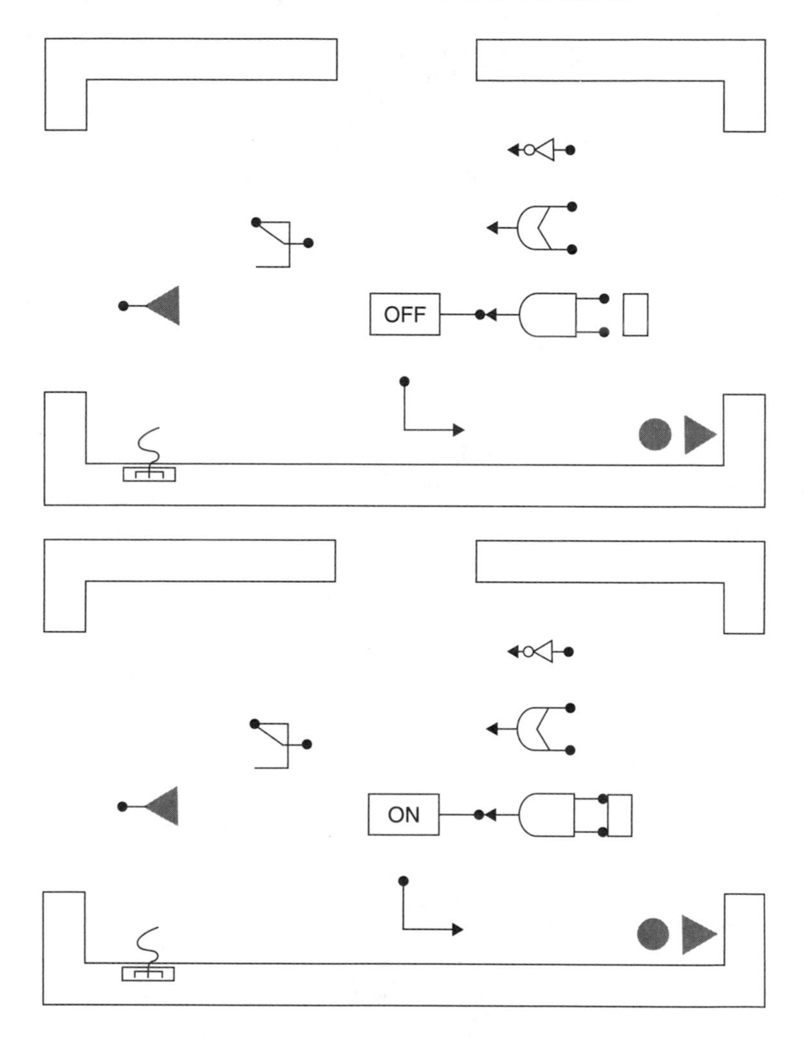

Fig 2.2. Two screens from the pretest–postest of scientific–technical discovery. Shaded areas, which were orange in the actual displays, represent the flow of power. The sequence of screens shows an "and-gate" being turned on. An "and-gate" is activated when two input nodes are simultaneously touched by the power source. An "and-gate" contrasts logically with an "or-gate," which can be activated when either one or the other input node is touched by the power source.

1990a, 1990b; Roberts, 1984). Consequently, repeated practice with games and other computer applications may enhance selected spatial skills.

Spatial representation is required by many, if not all, action video games. The action takes place in a virtual space that is shown one screen or one shot at a time. In order to play most games, one must develop a mental

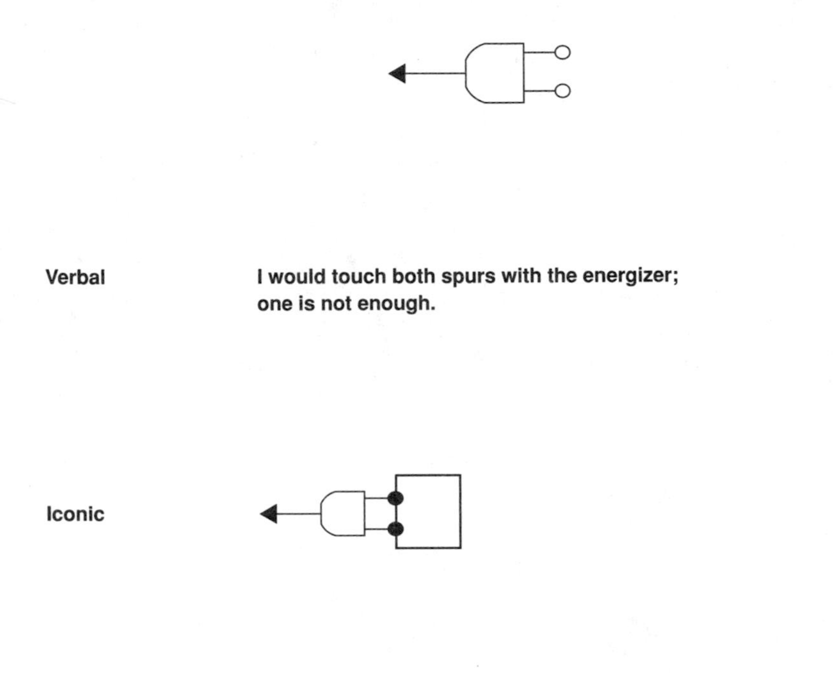

Fig 2.3. Different modes of representation used to answer pre- and posttest questions.

representation of the whole space and understand how each screen relates spatially to other parts of the space shown on different screens. One example is Castle Wolfenstein, a maze game in which a prisoner tries to escape from the castle, a Nazi prison; the prison is represented as a series of linked mazes. Each maze, in turn, represents a room in the castle; rooms are linked by virtual doorways into floors; floors are linked by virtual stairways into various levels of the castle. In the initial version in the 1980s, Castle Wolfenstein was represented as a series of two-dimensional mazes; later, it was represented as a series of three-dimensional mazes. In both cases, the principle is the same: To play effectively, one must figure out how a maze shown on one screen relates to mazes shown on other screens. In other words, in order to escape,

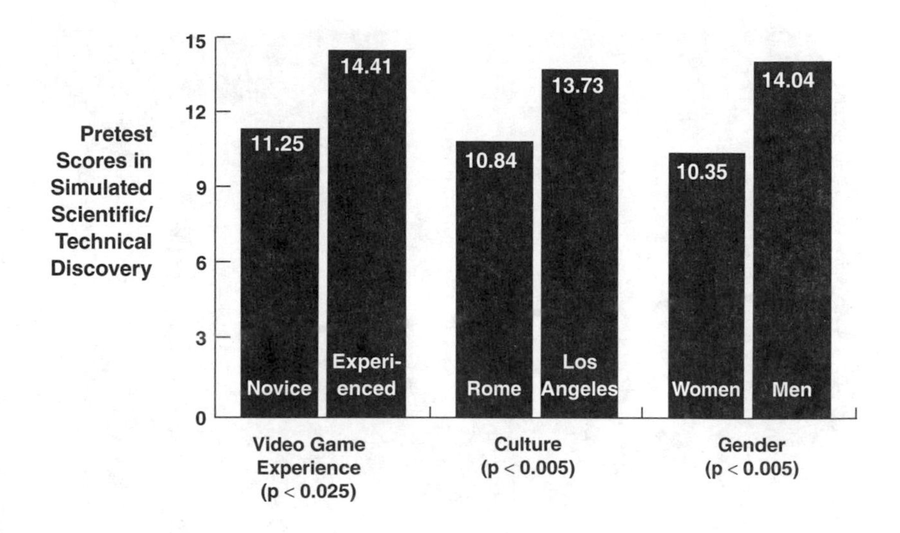

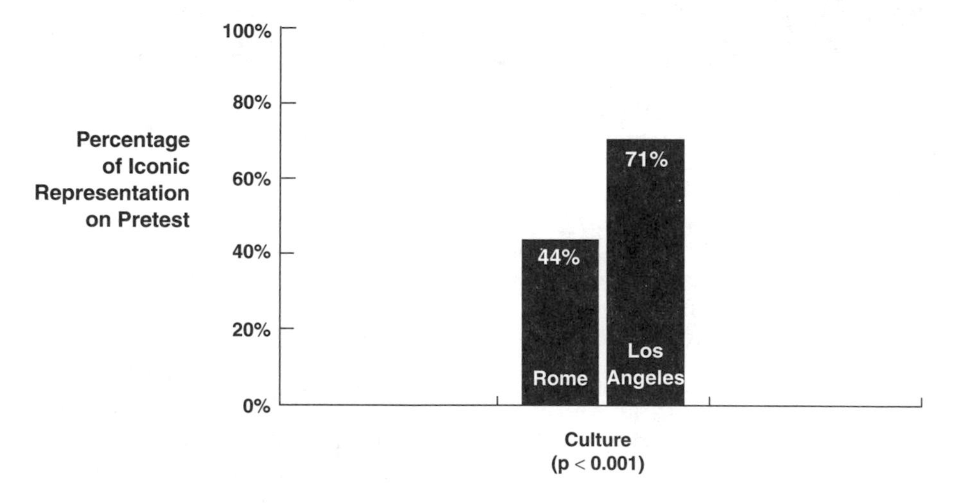

Fig 2.4. Significant long-term influences on simulated scientific-technical discovery and mode of representation.

one must figure out the layout of the castle. As a player becomes more expert in the game, he or she develops a spatial representation of the castle. Fig. 2.5 shows how this spatial representation expanded and elaborated as one player, age 15, gained more experience in the play. Note that the representation includes not just the castle, but also the location of various key objects within the castle. This example illustrates how skill with the technology of an action video game both requires and provides practice in

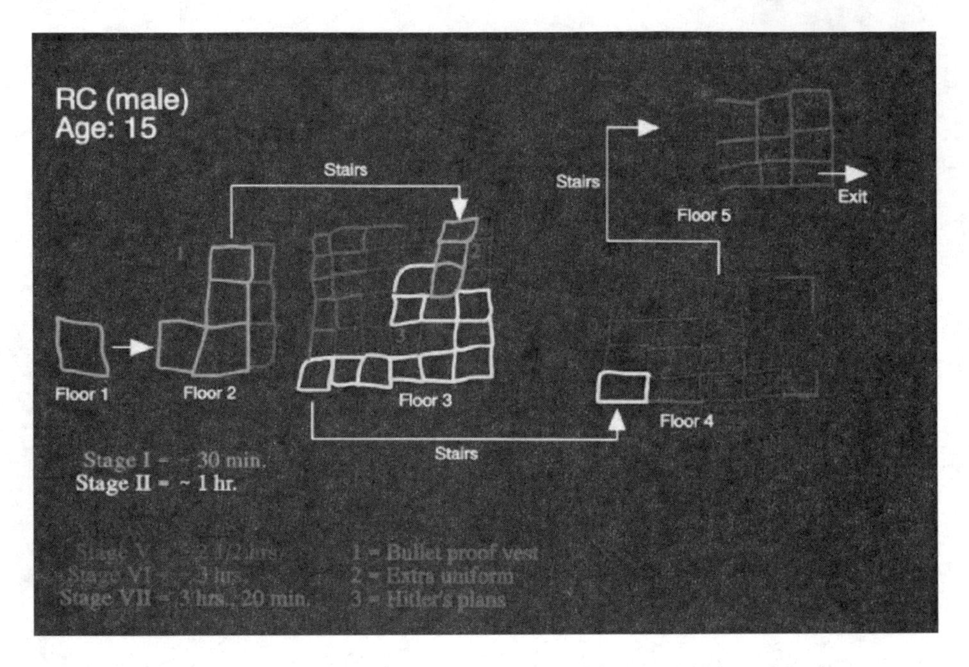

Fig 2.5. The progressive development of a spatial representation of Castle Wolfenstein over the first 3 hours and 20 minutes of play.

spatial representation. The external representations of separate mazes are transformed as they are internalized as one unified spatial representation of the castle as a whole.

The available evidence suggests that action games may serve as informal cultural tools for improving spatial skills more generally (Greenfield, Brannon, & Lohr, 1996; Okagaki & Frensch, 1994; Subrahmanyam & Greenfield, 1994; Subrahmanyam et al., 2001). Along similar lines, McClurg and Chaille (1987) showed that playing video games enhanced the spatial ability to mentally rotate three-dimensional objects in fifth-, seventh-, and ninth-grade students. Miller and Kapel (1985) found a similar positive effect of video games on the rotation of two-dimensional objects in seventh and eighth graders.

In a study of 10½- to 11½-year-olds, Subrahmanyam and Greenfield (1994) found that practice on a computer game (Marble Madness) reliably improved spatial performance (e.g., anticipating targets, extrapolating spatial paths) compared to practice on a computerized word game called Conjecture. Marble Madness involves guiding a marble along a three-dimensional grid using a joystick—the player has to keep the marble on the path and prevent it from falling off and prevent being attacked by intruders.

Verbal Representation. Another equally important question concerns the impact of Internet use on verbal representational skills. At a very basic level, Internet use involves reading and navigating around Web sites. In addition, the popular Internet applications such as instant messaging, e-mail, bulletin boards, and chat rooms involve the use of writing. Thus, the frequent use of the Internet may have important consequences for verbal representational skills. Unlike the medium of television and video/computer games, the Internet involves reading print and its use may actually result in more reading than before, albeit reading in a different medium. Second, the writing involved in online discourse is different from that found in traditional forms of written discourse, such as books and magazines, in that it has the features of both oral and written language. This is especially true of the Internet applications that are used for communication such as e-mail and instant messaging. For instance, chat conversations consist of shorter, incomplete, and grammatically simple and often incorrect sentences (Herring, 1996). Novel abbreviations are also rife (Greenfield & Subrahmanyam, in press). A question for the future is what will be the cumulative impact of such online reading and writing on verbal representational skills and ultimately for cultural definitions of verbal intelligence.

Mental Transformation

Video game expertise appears to have an impact on the development of mental transformation skills, such as those used in mental paper-folding tasks. In mental paper folding, a two-dimensional stimulus is mentally transformed into a three-dimensional stimulus. Greenfield, Brannon, & Lohr (1994) studied 82 undergraduates to assess the relationship between expertise in a three-dimensional action arcade video game, The Empire Strikes Back, and the skill of mental transformation as assessed in a mental paper-folding test (see Fig. 2.6 for sample items from the test). Although they found that short-term video game practice had no effect on mental paper folding, they found that video game expertise, developed over the long term, had a beneficial effect on the spatial skill of mental paper folding.

Implications for Intelligence

It is quite clear that a number of the skills being enhanced by computer technologies are also part of our cultural definitions of intelligence. Indeed, selective increases in nonverbal or performance IQ scores in recent years may be related, in part, to the proliferation of computer technologies in the environment that has occurred during the same period of time (Flynn,

Below are drawings each representing a cube that has been "unfolded." Your task is to mentally refold each cube and determine which one of the sides will be touching the side marked by an arrow.

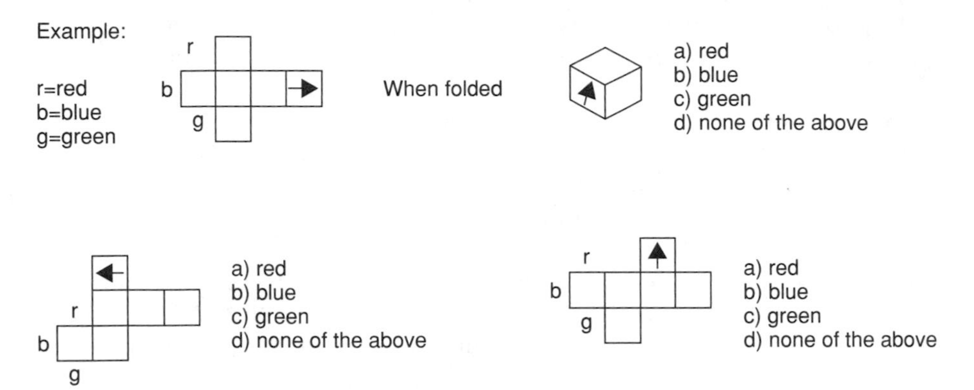

Fig 2.6. Sample item from the mental paper-folding test used in the study by Greenfield, Brannon, & Lohr (1994).

1994; Greenfield, 1998). Nonverbal or performance IQ tests and subtests are basically tests of different sorts of visual intelligence. For example, Chatters (1984) found that playing a video game exerted a significant positive effect on sixth-grade children's performance on the Block Design subtest of the Wechsler Intelligence Scale for Children (WISC). Similarly, Okagaki and Frensch (1994) reported that practice on the computer game Tetris (a game that requires the rapid rotation and placement of seven different-shaped blocks) significantly improved mental rotation time for undergraduates. However, note that mental rotation items are a traditional item type in ability tests. Indeed, in an early study, Gagnon (1985) found that 5 hours of practice on the action video game Targ improved performance on the mental rotation items of a vocational aptitude test. All of these findings contribute to our thesis that there is a tight relationship between a culture's technologies and its definitions of intelligence. They provide a clue that the changing modes of verbal representation seen on the Internet will ultimately lead to changed definitions of verbal intelligence.

WEAVING TECHNOLOGY AND COGNITIVE SKILLS

We now turn to a different kind of technology: Mayan backstrap loom weaving. It too has tight links with the three levels of cognition: attention, representation, and mental transformation. Although the levels of cognition are the same, the particular skills are different. As in the case of computers, the relevant skills are those that are engaged by this particular technology.

Attention

The transmission of weaving technology both utilizes and further socializes processes of visual attention (Greenfield, Brazelton, & Childs, 1989). Relative to Euro-American babies, Zinacantec Maya infants are born with extended visual attention spans (Brazelton, Robey, & Collier, 1969). Zinacantec caregivers then capitalize on this skill as they teach girls to weave (Maynard, Greenfield, & Childs, 1999). Learning by observation, which depends on extended focused visual attention, becomes extremely important as girls learn to weave (Greenfield, Brazelton, & Childs, 1989). A young girl must watch her mother or sister for months before trying weaving herself (Haviland, 1978). Even the first time girls try weaving themselves, they watch an expert model weave more than they weave themselves: First-time weavers spent 53% of their time observing the teacher, 39% weaving, and only 8% distracted (Childs & Greenfield, 1980). Learners in the United States, who have not received practice in the extended visual attention required by observational learning, can become very frustrated at having to watch so long before weaving themselves (Greenfield et al., 1989).

Representation

Facility with weaving technology influences strategies of visual representation (Greenfield & Childs, 1977; Greenfield, Maynard, & Childs, 2003). Greenfield and Childs (1977) found that unschooled girls who were weavers showed attention to the construction of cloth as they made thread-by-thread representations of striped woven textiles when given wooden sticks as a representational medium; unschooled boys of the same age (who were not weavers) did not create this type of representation. Not being practiced in weaving technology, they did not represent the actual construction of the textile designs; instead, these boys focused on how the textiles might look from a distance in their representations. However, formal schooling, for a small group of boys who received it, had the same effect as knowing how to weave. Schooled teenage boys also provided thread-by-thread analyses of the textile patterns in their representations. Apparently the cognitive tools provided by formal education were a substitute for weaving technology in developing this type of visual representation of textile patterns.

Other researchers also found a link between expertise in the use of weaving technology and skill in pattern representation. Comparing expert adolescent Navajo rug weavers to other Navajo who did not know how to weave, Rogoff and Gauvain (1984) found increased ability in representing familiar, but not novel patterns. In another kind of weaving, straw weaving in rural Northeast Brazil, Saxe and Gearhart (1990) found that experience with the psychological technology of knowing how to weave influenced representation of topological information in novel patterns. Knowing how to use a particular technology influences representation of spatially related information.

Mental Transformation

Experience with weaving technology also enhances relevant skills in spatial transformation. Maynard and Greenfield (2003) investigated the link between experience in weaving and the development of spatial transformation by examining mental transformations involved in creating the warp of a loom. Prior fieldwork produced a hypothesis that Zinacantec weaving tools are adapted to the developmental status of learners (Greenfield, 2000a, 2000b), specifically whether or not they have reached a stage where they are capable of mental transformation.

Most girls first learn to wind a warp on a toy loom (see Fig. 2.7), which is adapted to young girls, ages 3 to 5 or 6. Older girls, who usually have had some experience in weaving, wind on a warping frame (see Fig. 2.8). The winding tool that is adapted for older girls reflects an advanced stage of cognitive development, one that requires mental transformation. For

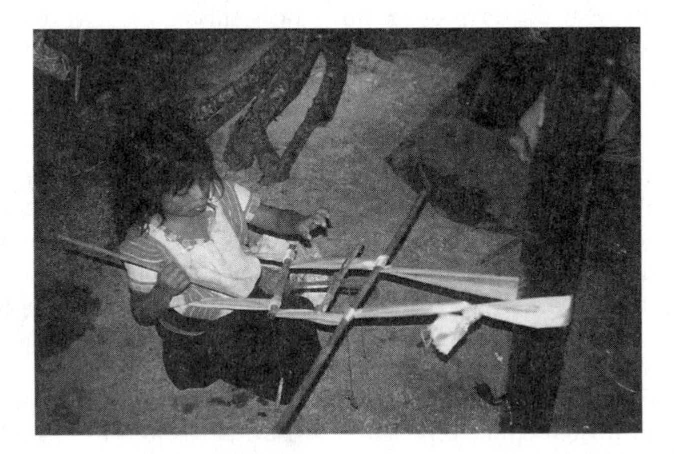

Fig 2.7. A toy loom. The weaver has wound her warp directly on the loom between the two end sticks. Photograph by Patricia Greenfield, Nabenchauk, 1991.

Fig 2.8. Warping frame, or *komen*, which requires mental transformation to visualize how the woven material will appear. A warp has already been wound. The left side of the threads will go to one end of the loom, say the top, while the threads on the right side will go to the other, the bottom.

example, one needs to understand that the left side of the threads in Fig. 2.8 will end up at one end of the loom, for example the top of the loom shown in Fig. 2.7, while the right side of the threads in Fig. 2.8 will end up at the other end of the loom, for example, the bottom of the loom shown in Fig. 2.7. Once this transformation is carried out, either mentally or in practice, one implication is that the resulting piece of cloth will be approximately twice as long as its length on the warping frame, where it is in essence folded in half.

The winding tool that is adapted to younger girls reflects a less advanced stage of cognitive development because it does not require mental transformation (Piaget & Inhelder, 1956). The weaver simply winds the warp in figure eights from top to bottom; the length of the resulting cloth matches the length between the top and bottom sticks. The relationship is one of perceptual matching rather than mental transformation. In contrast, the warping frame requires an ability to perform mental transformations; mental transformations are required to predict what the cloth will look like once it is woven. We created two types of tasks related to weaving: the toy loom tasks (an example is presented in Fig. 2.9) and the warping frame tasks (an example is presented in Fig. 2.10).

It was predicted that children over the age of 6 would be able to perform the mental transformations involved in understanding the warping frame if they had had some experience in weaving. Children with experience in weaving were Zinacantec girls, whereas children with no experience were Zinacantec boys and American children.

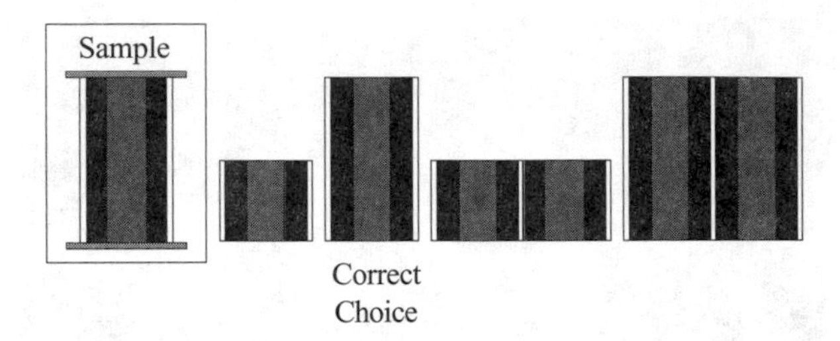

Fig 2.9. An example of a loom with four choices. This is a direct perceptual-matching task. The second choice to the right of the loom is what the warp will look like when woven.

A crossover task was designed to measure transformation abilities cross-culturally as well as to examine domain transfer across the two types of tasks. The crossover task, a more familiar task type in the United States, was referred to as the "knots" tasks, based on work by Piaget and Inhelder (1956). The knots were loops of string ("necklaces") with spools of different-colored thread on them. We turned the first loop of each set into a figure eight, thus, creating a situation that requires mental transformation to predict what the configuration of the spools will be once the knot or figure eight is unlooped. An example is presented in Fig. 2.11.

Participants in Los Angeles and the Zinacantec Maya community of Nabenchauk, ages 4 through 13, were asked to perform match-to-sample tasks of three different types: the toy loom, the warping frame, and the knots tasks.

Zinacantec girls performed significantly better on the warping frame tasks than did the Zinacantec boys or American children of either sex. This pattern of results demonstrates that only direct experience with weaving technology has an impact, not the passive familiarity experienced by Zinacantec

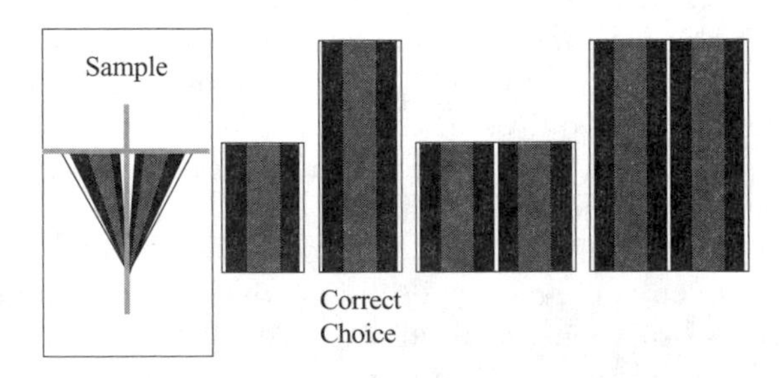

Fig 2.10. An example of a *komen* with four choices. This is a task requiring mental transformation. The correct answer is the second choice to the right of the *komen*.

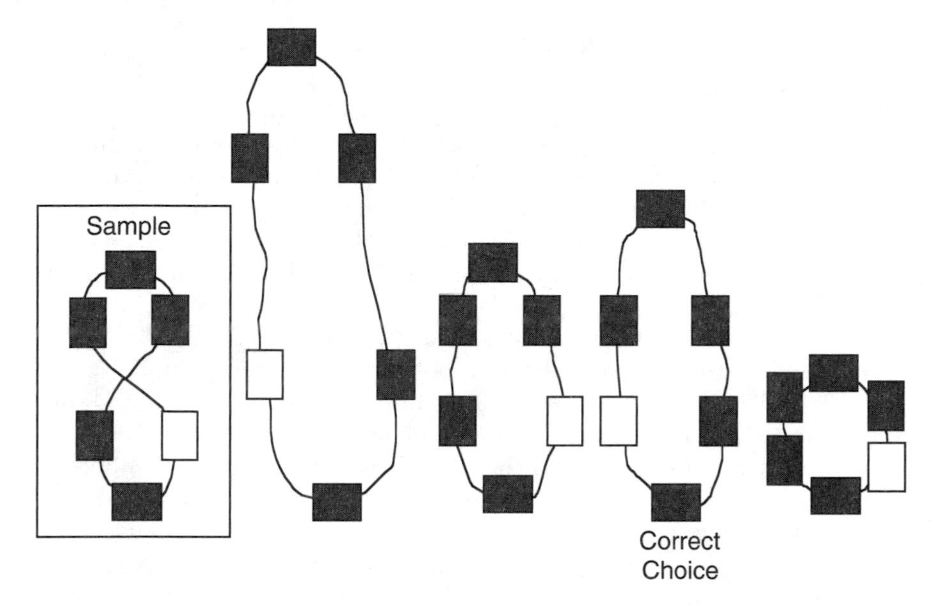

Fig 2.11. An example of a knots task with four choices. This is a task requiring mental transformation. The correct answer is the third choice to the right of the figure eight.

boys who have seen, but not used, the technology. Although children in Los Angeles performed less well on mental transformation in the context of the warping frame problems, they performed better on the knots tasks, a task whose structure was probably more familiar to these schooled children. These results indicate that a particular technology can develop mental operations that are tailored to a particular domain; the operations then take on a domain-specific form. These mental operations, as they are used in the culturally relevant domain, we hypothesize, will then be considered part of a cultural definition of technological intelligence.

At the same time, Zinacantec weaving apprenticeship also respects stages in the development of mental transformation skills. We found that Zinacantec girls begin to wind a warp on a warping frame at the age at which they have the requisite level of cognitive development. Analysis of patterns of children's answers revealed that children were using perceptual matching at the age (less than 6) when they would be winding a warp on a toy loom; they then switched to mental transformation at the age (over 6) when they would be winding a warp on the warping frame. These results indicate that weaving apprenticeship not only develops relevant skills in spatial transformation, it also respects the developmental timetable by which those skills develop.

The Adoption of New Symbolic Tools

The nature of symbolic tools used by a particular group of people also changes over time. Although the backstrap loom, of ancient Maya origin, has remained the same, there are new symbolic tools used in the production of textiles. Specifically, paper patterns designed for embroidery are now being used by the Zinacantecs. Greenfield (1999) labels the paper patterns metarepresentational because they are tools for creating patterns, that is for creating other representations. Whereas there were no metarepresentational tools in Zinacantán up through the 1970s, Zinacantec females began using paper patterns designed for embroidery to weave sometime in the late 1980s. The technology was not indigenous to Zinacantec culture, but rather imported from Mexican culture and adapted to weaving. The use of the cross-stitch patterns for embroidery relies on perceptual matching, as there is a one-to-one relationship between the grid on the paper and what is to be embroidered onto cloth. However, Zinacantec girls began using the paper cross-stitch patterns to weave, a task which required a mental transformation because weaving is not done in squares. The conversion was one square to one warp thread. However, a one-to-one ratio could not be used in the cross-wise or weft thread dimension. Some weavers transformed the length of each square in the pattern to four weft threads. Zinacantec girls had appropriated a new symbolic tool, the printed pattern, and transformed it, as part of the process of cultural appropriation (Saxe, 1999). In so doing, they revealed some skills in mental transformation.

CULTURAL VALUES AND THE USE OF CULTURAL TOOLS

The development of intelligence is influenced by cultural values that apply relatively greater emphasis to technological or social intelligence (Mundy-Castle, 1974). Zinacantec weaving is connected to their ethnotheory of development that implies that a girl will weave when she has enough soul, meaning that she can listen to instruction, follow instructions, do what is needed, and tolerate frustration (Zambrano & Greenfield, 2004). Relatedly, weaving is not valued as a technical skill; rather, it is valued for its social aspects: the social and interactional aspects of the learning process, the social utility of what is woven, and the enhancement of a girl's marriageability by being a skilled weaver. Whereas we have been focusing on the role of weaving in developing particular forms of technological intelligence, the Zinacantecs have traditionally been much more focused on weaving's role in reflecting and developing social and emotional intelligence. This focus on social and emotional intelligence contrasts with American attitudes toward computer technology, where developing technological intelligence is of primary importance. A video game has, by definition, no external social goal or purpose, whereas in Zinacantán weaving does. In addition, electronic games are often played in an individual or private setting, whereas weaving is generally done in a social setting—the family courtyard. Finally, weaving apprenticeship depends more heavily on interaction with others than does learning how to play an electronic game. When playing a video game, one is also playing with or against virtual, rather than real, people, and, in the more recent multiplayer online games, one is playing with or against real people, but people who are anonymous and disembodied. For all these reasons, video and computer games might foster technological intelligence at the expense of social intelligence. This would be less likely in Zinacantán where weaving is social in function, setting, and mode of apprenticeship, although equally technological in its execution.

CONCLUSIONS

In this chapter we reported the ways in which different technologies develop intelligence on the three cognitive levels of attention, representation, and mental operations. Though our focus was on two specific technologies about which there has been a lot of research, there are many other findings demonstrating the effects of technology on the culturally valued cognitive skills that constitute intelligence. For example, on the level of visual representation, Stigler (1984) showed that expert abacus users develop a mental representation of an abacus and make errors reflecting that representation when asked to perform mental calculations.

Additionally, we discussed the specific tools of computer technology and weaving. The skills developed by those cultural tools are very different ones. Because a culture's technologies determine that culture's view of technological intelligence, then, to the extent that one has different technologies, one is going to have different definitions of technological intelligence. Tools differ across different ecological niches; thus, the forms of intelligence that may develop vary also. For example, the divided attention that is useful in handling video games and the Internet would be anathema to the Zinacantecs, who favor the undivided visual attention skills that are useful in learning to weave.

At the same time, tools and their ecological settings are not constant even within a single cultural context. The pan-human capacity to invent and use tools leads to adaptation of those tools to new cultural places or activities. As the environment changes, the nature and function of tools may change as well. Will new verbal conventions developed in adaptation to the Internet lead to new definitions of verbal intelligence in the United States? Will the use of paper patterns in weaving lead to new definitions of technological intelligence in Zinacantán? We predict that both of these changes will occur. Just as technologies are not static, neither are cultural definitions of technological intelligence; the two, by their very nature, must evolve together.

The movement from subsistence to commerce in Zinacantán has already begun to transform weaving apprenticeship from a socially guided process to one involving more individual experimentation and discovery (Greenfield, 1999; Greenfield, Maynard, & Childs, 2003). Will the accelerating movement from subsistence to commerce in Zinacantán change the emphasis from weaving as a set of social skills to textile production as a set of technological skills with commercial value? On a trip to Nabenchauk just a few weeks before this chapter was completed, Greenfield noted the beginnings of this transition. Not only the nature of technological intelligence but also its relative social importance are both culturally variable and historically contingent.

REFERENCES

Boyd, R., & Silk, J. B. (2000). *How humans evolved.* New York: Norton.

Brazelton, T. B., Robey, J. S., & Collier, G. (1969). Infant development in the Zinacanteco indians of Southern Mexico. *Pediatrics, 44,* 274–283.

Bruner, J. S. (1966). On cognitive growth. In J. S. Bruner, R. R. Olver, & P. M. Greenfield (Eds.), *Studies in cognitive growth* (pp. 1–67). New York: Wiley.

Chatters, L. B. (1984). *An assessment of the effects of video game practice on the visual motor perceptual skills of sixth grade children.* Unpublished doctoral dissertation, University of Toledo, Ohio.

Childs, C. P., & Greenfield, P. M. (1980). Informal modes of learning and teaching: The case of Zinacantec weaving. In N. Warren (Ed.), *Studies in cross-cultural psychology, Vol. 2* (pp. 269–316.) London: Academic Press.

Cole, M., & Griffin, P. (1980). Cultural amplifiers reconsidered. In D. R. Olson (Ed.), *The social foundations of language and thought* (pp. 343–364). New York: Norton.

Dasen, P. (1984). The cross-cultural study of intelligence: Piaget and the Baolé. In P. S. Fry (Ed.), *Changing conceptions of intelligence and intellectual functioning: Current theory and research* (pp. 107–134). Amsterdam: Elsevier Science.

Flynn, J. R. (1994). IQ gains over time. In R. J. Sternberg (Ed.), *Encyclopaedia of human intelligence* (pp. 617–623). New York: Macmillan.

Gagnon, D. (1985). Videogames and spatial skills: An exploratory study. *Educational Communication and Technology Journal, 33,* 263–275.

Gomez, L. M., Bowers, C., & Egan, D. E. (1982). Learner characteristics that predict success in using a text-editor tutorial. In *Proceedings of Human Factors in Computer Systems* (pp. 176–181). New York: ACM Press.

Green, C. S., & Bavelier, D. (2003). Action video game modifies visual selective attention. *Nature, 423,* 534–537.

Greenfield, P. M. (1983). Video games and cognitive skills. In *Video games and human development: Research agenda for the '80s* (pp. 19–24). Cambridge, MA: Monroe C. Gutman Library, Graduate School of Education.

Greenfield, P. M. (1984a). *Mind and media: The effects of television, video games and computers.* Cambridge, MA: Harvard University Press.

Greenfield, P. M. (1984b). A theory of the teacher in the learning activities of everyday life. In B. Rogoff & J. Lave (Eds.), *Everyday cognition: Its development in social context* (pp. 117–138). Cambridge, MA: Harvard University Press.

Greenfield, P. M. (1985). Multimedia education: Why print isn't always best. *American Educator, 9*(3), 18–21, 36, 38.

Greenfield, P. M. (1987). Electronic technologies, education, and cognitive development. In D. E. Berger, K. Pezdek, & W. P. Banks (Eds.), *Applications of cognitive psychology* (pp. 17–32). Hillsdale, NJ: Lawrene Erlbaum Associates.

Greenfield, P. M. (1990a). Video games as tools of cognitive socialization. *Psicologia Italiana, 10*(1), 38–48.

Greenfield, P. M. (1990b). Video screens: Are they changing how children learn? *Harvard Educational Letter, 6*(2), 1–4.

Greenfield, P. M. (1998). The cultural evolution of IQ. In U. Neisser (Ed.), *The rising curve: Long term gains in IQ and related measures* (pp. 81–123). Washington, DC: American Psychological Association.

Greenfield, P. M. (1999). Cultural change and human development. In E. Turiel (Ed.), *Development and cultural change: Reciprocal processes. New directions for child and adolescent development* (pp. 37–59). San Francisco: Jossey-Bass.

Greenfield, P. M. (2000a). Children, material culture, and weaving: Historical change and developmental change. In J. S. Deeverenski (Ed.), *Children and material culture* (pp. 72–86). London: Routledge.

Greenfield, P. M. (2000b). Culture and universals: Integrating social and cognitive development. In L. P. Nucci, G. B. Saxe, & E. Turiel (Eds.), *Culture, thought, and development* (pp. 231–277). Mahwah, NJ: Lawrence Erlbaum Associates.

Greenfield, P. M. (2004). *Weaving generations together: Evolving creativity in the Maya of Chiapas.* Santa Fe, NM: SAR Press.

Greenfield, P. M., Brannon, C., & Lohr, D. (1996). Two-dimensional representation of movement through three-dimensional space: The role of video game expertise. *Journal of Applied Developmental Psychology, 15*, 87–103.

Greenfield, P. M., Brazelton, T. B., & Childs, C. P. (1989). From birth to maturity in Zinacantan: Ontogenesis in cultural context. In V. Bricker & G. Gossen (Eds.), *Ethnographic encounters in Southern Mesoamerican: Celebratory essays in honor of Evon Z. Vogt* (pp. 177–356). Albany, NY: Institute of Mesoamerican Studies, State University of New York.

Greenfield, P. M., Camaioni, L., Ercolani, P., Weiss, L., Lauber, B., & Perucchini, P. (1994). Cognitive socialization by computer games in two cultures: Inductive discovery or mastery of an iconic code? *Journal of Applied Developmental Psychology, 15*, 59–85.

Greenfield, P. M., & Childs, C. P. (1977). Weaving, color terms, and pattern representation: Cultural influences and cognitive development among the Zinacantecos of Southern Mexico. *Interamerican Journal of Psychology, 11*, 23–48.

Greenfield, P. M., deWinstanley, P., Kilpatrick, H., & Kaye, D. (1994). Action video games and informal education: Effects on strategies for dividing visual attention. *Journal of Applied Developmental Psychology, 15*, 105–123.

Greenfield, P. M., Keller, H., Fuligni, A., & Maynard, A. (2003). Cultural pathways through universal development. *Annual Review of Psychology, 54*, 461–490.

Greenfield, P. M., Maynard, A. E., & Childs, C. P. (2003). Historical change, cultural apprenticeship, and cognitive representation in Zinacantec Maya children. *Cognitive Development, 18*, 455–487.

Greenfield, P. M., & Subrahmanyam, K. (2003). Online discourse in a teen chatroom: New codes and new modes of coherence in a visual medium. *Journal of Applied Developmental Psychologye 24*(6), 713–738.

Gross, E. (2003, June). *Adolescents and the Internet: Myth and reality.* Paper presented at the annual meeting of the Jean Piaget Society for Knowledge and Development, Chicago.

Guberman, S. R. (1996). The development of everyday mathematics in Brazilian children with limited formal education. *Child Development, 67*(4), 1609–1623.

Haviland, L. K. M. (1978). *The social relations of work in a peasant community.* Unpublished doctoral dissertation, Harvard University, Cambridge, Massachusetts.

Herring, S. C. (1996). Introduction. In S. Herring (Ed.), *Computer-mediated communication: Linguistic, social, and cross-cultural perspectives* (pp. 1–12). Philadelphia: Benjamins.

Lave, J. (1977). Tailor-made experiments and evaluating the intellectual consequences of apprenticeship training. *Anthropology and Education Quarterly, 8,* 177–180.

LeVine, R. A. (2002, October). *Cultural priming and precocity in infancy and early childhood.* Lecture presented at FPR-UCLA Center for Culture, Brain, and Development, Los Angeles.

Maynard, A. E., & Greenfield, P. M. (2003). Implicit cognitive development in cultural tools and children: Lessons from Mayan Mexico. *Cognitive Development, 18,* 489–510.

Maynard, A. E., Greenfield, P. M., & Childs, C. P. (1999). Culture, history, biology, and body: Native and non-native acquisition of technological skill. *Ethos, 27*(3), 379–402.

McClurg, P. A., & Chaille, C. (1987). Computer games: Environments for developing spatial cognition? *Journal of Educational Computing Research, 3,* 95–111.

Miller, G. G., & Kapel, D. E. (1985). Can non-verbal, puzzle type microcomputer software affect spatial discrimination and sequential thinking skills of 7th and 8th graders? *Education, 106,* 160–167.

Mundy-Castle, A. C. (1974). Social and technological intelligence in Western and non-Western cultures. *Universitas, 4,* 46–52.

Nunes, T., Schliemann, A. D., & Carraher, D. W. (1993). *Street mathematics and school mathematics.* Cambridge, UK: Cambridge University Press.

Okagaki, L., & Frensch, P. A. (1994). Effects of video game playing on measures of spatial performance: Gender effects in late adolescence. *Journal of Applied Developmental Psychology, 15,* 33–58.

Olson, D. R., & Bruner, J. S. (1974). Learning through experience and learning through media. In D. E. Olson (Ed.), *Media and symbols: The forms of expression, communication, and education* (pp. 125–150). Chicago: University of Chicago Press.

Pellegrino, J. W., & Kail, R. (1982). Process analyses of spatial aptitude. In R. J. Sternberg (Ed.), *Advances in the psychology of human intelligence, Vol. 1* (pp. 311–365). Hillsdale, NJ: Lawrence Erlbaum Associates.

Piaget, J., & Inhelder, B. (1956). *The child's conception of space.* London: Routledge and Kegan Paul.

Posner, M. I., Snyder, C. R., & Davidson, B. J. (1980). Attention and the detection of signals. *Journal of Experimental Psychology, 109,* 160–174.

Price-Williams, D. R., Gordon, W., & Ramirez, M., III. (1969). Skill and conservation: A study of pottery-making children. *Developmental Psychology, 1,* 769.

Roberts, R. (1984, April). *The role of prior knowledge in learning computer programming.* Paper presented at the annual meeting of the Western Psychological Association, Los Angeles, CA.

Rogoff, B., & Gauvain, M. (1984). The cognitive consequences of specific experiences: Weaving versus schooling among the Navajo. *Journal of Cross-Cultural Psychology, 15*(4), 453–475.

Salomon, G. (1988). AI in reverse: Computer tools that turn cognitive. *Journal of Educational Computing Research, 4*(2), 123–139.

Saxe, G. B. (1994). Studying cognitive development in sociocultural context: The development of a practice-based approach. *Mind, Culture, and Activity, 1*(3), 135–157.

Saxe, G. B. (1999). Cognition, development, and cultural practices. In E. Turiel (Ed.), *Development and cultural change: Reciprocal processes. New directions for child and adolescent development* (pp. 19–35). San Francisco: Jossey-Bass.

Saxe, G. B., & Gearhart, M. (1990). A developmental analysis of everyday topology in unschooled straw weavers. *British Journal of Developmental Psychology, 8,* 251–258.

Saxe, G. B., & Moylan, T. (1982). The development of measurement operations among the Oksapmin of Papua New Guinea. *Child Development, 53,* 1242–1248.

Scheibel, A. (1996, March). Introduction to Symposium on the Evolution of Intelligence. Center for the Study of Evolution and the Origin of Life, University of California, Los Angeles.

Sternberg, R. J., Conway, B. E., Ketron, J. L., & Bernstein, M. (1981). People's conceptions of intelligence. *Journal of Personality and Social Psychology, 4,* 37–55.

Stigler, J. W. (1984). "Mental abacus": The effect of abacus training on Chinese children's mental calculation. *Cognitive Psychology, 16,* 146–176.

Subrahmanyam, K., & Greenfield, P. M. (1994). Effect of video game practice on spatial skills in girls and boys. *Journal of Applied Developmental Psychology, 15,* 13–32. Also In P. M. Greenfield and R. Cocking (Eds.), *Interacting with video* (pp. 95–114). Norwood, NJ: Ablex.

Subrahmanyam, K., Greenfield, P. M., Kraut, R., & Gross, E. (2001). The impact of computer use on children's development. *Journal of Applied Developmental Psychology, 22,* 7–30.

Tikhomirov, O. K. (1974). Man and computer: The impact of computer technology on the development of psychological processes. In D. E. Olson (Ed.), *Media and symbols: The forms of expression, communication, and education* (pp. 357–382). Chicago: University of Chicago Press.

Vanderwater, E., Wartella, E., & Rideout, V. (2003). *From Zero to Six.* San Francisco: Kaiser Family Foundation.

Vygotsky, L. S. (1962). *Thought and language.* Cambridge, MA: MIT Press.

Vygotsky, L. S. (1978). *Mind in society: The development of higher psychological processes.* Cambridge, MA: Harvard University Press.

Wober, M. (1974). Towards an understanding of the Kiganda concept of intelligence. In J. W. Berry and P. R. Dasen (Eds.), *Culture and cognition: Readings in cross cultural psychology* (pp. 261–280). London: Methuen.

Zambrano, I., & Greenfield, P. (2004). Ethnoepistemologies at home and at school. In E. Grigorenko and R. Sternberg (Eds.), *Culture and competence* (pp. 251–272). Washington, DC: American Psychological Association.

3

Technology and Intelligence in a Literate Society

David R. Olson
Ontario Institute for Studies in Education
University of Toronto

The goal of the cognitive sciences is to explain our most valued, and distinctively human, competencies that make up our intelligence. The problem is how to best construe those competencies in such a way as to explain both what is distinctively human about those abilities and the special shape they take in a modern, bureaucratic, technological society. The solution it avers is that intelligence is misrepresented if seen only as a basic, largely innate, trait or disposition. In a modern society, it would be more correctly represented as a metarepresentation or rationalization of those more basic processes. These metarepresentational systems are heavily dependent on culture in general and literacy and schooling in particular.

Although the concept of intelligence appears to be a unitary and universal one, the concept functions in at least three quite different ways. It is a central notion in the folk psychology we all use to talk about our children and our friends. Although these folk notions are misleading, one would be foolhardy indeed to legislate the use of the term. Second, it is used to label a disposition or trait in *differential* or individual-differences psychology and in population-genetic studies of heritability. Third, it is used as a general term to describe the basic cognitive competencies involved in learning language and mathematics as well as the more specialized arts and sciences. It is this third view that has sponsored the notion of modularity of mind as well as much recent cognitive and developmental research. It is this third view of intelligence that addresses the psychological life of persons as conscious intentional agents affected by development and capable of learning.

It is important to distinguish the second and the third conceptions of intelligence. Is intelligence a biologically based trait that is manifested in levels of acquired abilities? Or is it those abilities themselves? For Binet, and for individual-differences or trait theorists since, intelligence was what explained differences in school achievement, or in Binet's terms, a person's ability to benefit from schooling (see also Seigler, 1992). For Piaget (1950), who used some of the very test items invented by Binet, intelligence was not a disposition or trait but rather the very mental activities and achievements that were involved in solving those problems. These activities he labeled "the psychology of intelligence." That is, intelligence was what developed rather than some underlying disposition that allowed that competence to develop.

The two conceptions rest on quite different methodologies. Binet could rank performance relative to that of other persons, that is, relative to the statistical norms of the distribution or population. These traits or dispositions, including measured intelligence, IQ, aspire to objectivity by summarizing across the intentional states, the beliefs, desires, and intentions of a person, in order to arrive at an underlying "causal" trait that somehow explains those intentional states. This invariant trait may then be related to other kinds of performance, as when IQ predicts (that is, accounts for) some of the variability in reading achievement. One problem is that despite a century of such research there is still no way of determining whether observed correlations across tasks indicate some underlying causal property—a mental trait that could be used to explain performances on a variety of tasks—or merely that the tasks share some property. Put another way, there is no established way to show that the tests are symptoms of an underlying causal trait or merely correlations among a variety of symptoms. Again, it is not clear that intelligence as an underlying trait exists, let alone explains anything. There may be no "there" there as Gertrude Stein famously said when, on the occasion of her first visit, she was asked of her impression of Oakland.

The same story could be told by reference to the much studied relation between ability and achievement tests. Ability tests tend to predict the extent to which one is likely to benefit from instruction. Again, it is not known if that is because they assess some underlying ability or simply indicate the extent to which one already possesses the relevant knowledge. Consequently, there is some interest in assigning greater weight to achievement tests than to so-called tests of ability, such as the SAT, as criteria for university admission (Cloud, 2001; Olson, 2003). This shift was anticipated in the educational reforms at the beginning of the 19th century when the goal of training the intellect—advanced through the study of classical languages and abstract mathematics—was rejected in favor of learning specialized bodies of knowledge as represented by the sciences. The newer assumption seems to have been that thinking will be improved not as a general ability but as a function of advanced forms of specialized knowledge. This entirely plausible assumption has dominated educational practice for well over a century.

The shift in social policy from attention to a presumed ability to specific achievement in a domain is also seen in personnel selection. Intelligence tests gained much of their prominence when they were found to be quite useful in allocating personnel to positions by the U.S. military in World War I; by World War II, education was sufficiently widespread that earned credentials such as high school certificate or a university degree were much more important in placing recruits. If one can know only one thing about a person, an estimate of general ability is useful; if more information is available, specific achieved experience and credentials are more important. Personnel are assigned to positions in modern bureaucracies on the basis of their credentials, their diplomas, and their experience, rather than on some general estimate of intelligence or creativity.

Schools have been less clear on the criteria used for promotion or access to advanced programs. In the general case, a child gains access to advanced programs on the basis of success with more elementary ones. However, such categories as special education and giftedness programs appeal to measures of ability rather than achievement. Universities sometimes use measures of ability to override a poor academic record. These procedures touch on an important moral issue, that of justice. Achievements are the products of intentional actions for which one has taken responsibility and thereby earned a right or entitlement. To confer benefits on the basis of ability is to gift entitlements without the recipient having done anything to earn them. This would seem to be an injustice; in a just system one earns rewards, they are not gifted by race, class, or genes.

The uncertain relations between measures of ability and school achievement are all the more problematical when they are used to interpret school achievement and school failure. This is because the relations that are obtained in studies of populations are applied inappropriately in the individual case. That is, if one scores poorly on an item in an ability test, it is impossible to infer with any certainty that this failure is the result of a more general cognitive disability or more simply of the lack of the appropriate beliefs and desires. If one fares poorly on a test, is it because of a lack of ability or a failure to do one's homework? Consequently, predictions of success in individual cases are surprisingly unimpressive.

With the help of Renata Valtin and Oliver Thiel of Humboldt University of Berlin, we calculated the likelihood of predicting superior performance on an achievement test on the basis of a previously assessed ability test. The correlation was close to that obtained in hundreds of such studies, in this case $r = 0.51$. About two thirds of the high-ability children scored in the top half of the class, whereas about one third of the average-ability children did. However, because the high-ability group was only 10% of the population, more than twice as many nongifted children achieved top scores. If a teacher were looking for excellence among those labeled gifted, he or she would overlook the majority of good performances! Even among those labeled gifted, only some do well, *and it is impossible to predict who those will be.* Better, I suggest, to stick to the use of actual achievements as a basis of awarding privileges and credentials.

Geneticists such as Lewontin (1976) and psychologists such as Coyne (2000, p. 13) tell us that discourse about any individual person's intelligence, that is, their performance on some challenging intellectual task, often confuses valid claims made by population geneticists about *heritability,* a technical notion, with more popular notions of *inheritability,* the effects of an individual's genes on his or her behavior. Heritability can be measured: For a given trait in a given population, it is the proportion of the total observed variation among individuals attributable to variation in their genes. Even located genes account for very little of that variability. Researchers who located the ELAC2 gene for prostate cancer say that variations in the gene "are probably only responsible for two to five percent of cases of the disease" (Easton, 2001, p. 3). Inheritability, the role of one's genes in regulating any activity, is impossible to determine. In the case of any superb performance, it is unclear if the performance is to be traced to the genes one received from one's mother or the advice one received. Tests of intelligence rank a person's performance relative to a population. What they cannot do is indicate what that person was doing, thought he was doing, was trying to do, or how he went about doing it. That is, such tests cannot address the question of intentionality.

This fact was recognized early on by John Dewey, whose Progressivism has been referred to as "America's gift to the wider educational world" (Gardner, 2001, p. 128), and by Jean Piaget, who offered a more extensive and empirical theory of intelligence and a new methodology. Whereas Dewey (1972) insisted that agency and intentionality were central features of learning and thinking, Jean Piaget (1950) is more widely credited with changing the question about intelligence from "who has (more of) it?", the question that has occupied test-makers for over a century, to "what is it?" Whereas for Binet and his followers in the intelligence testing tradition, a test item was chosen so that it would discriminate the more from the less able, for Piaget the question became that of attempting to determine just what children or adults were doing or thought they were doing that would give rise to their particular performances. Younger children's inadequate performances were recognized, for the first time, not as simple failures, but as important steps toward adult-like solutions. Cognitive studies for the past 40 years have generally been advanced in this quasi-Piagetian tradition, that is, in the tradition of attempting to spell out what subjects know, want, and attempt, and the procedures they adopt to achieve their goals. These procedures are ideally spelled out in terms of a set of cognitive processes or strategies, most of which can be taught.

I can illustrate this shift in perspective on intelligence by reference to some of my own early research (Olson, 1970) on children's ability to reconstruct a diagonal pattern on a checker board, a task routinely failed by children under 5 or 6 years of age. The task is similar in form to the hundreds of items that make up IQ tests used to rank children in ability. Under the influence of Piaget as well as of Bruner (1960), the question was changed from "who has more ability" to "what are they doing?" This question could be asked of both those who succeeded and those who failed. The question was one of determining experimentally what mental representation of the event children were using and the conditions under which they would shift to a more appropriate representation.

The empirical finding was that younger children began their diagonals in the appropriate corner but were soon diverted into either a vertical column or a horizontal row on the board. The explanation I ultimately arrived at (with a suggestion from Janellen Huttenlocher) was that the children's actions were premised on adjacency, that is, they placed a checker on the nearest square rather than the more distant one making up the diagonal. Even failing children, I was able to show, had noticed that the checkers in the diagonal were in some sense next to each other in that they formed a straight line. But they were caught by Pythagorean fact that the hypotenuse is necessarily longer than the sides of a square. The diagonal falls along this hypotenuse. If children use adjacency to determine their placement of the checker, they will place it on a row or a column rather than on the diagonal. When they did begin to perform correctly they did so by redescribing the diagonal in terms of the properties of the rows and columns. Either they explicitly excluded the nearest squares, that is, excluding the rows or columns, or they invented a higher order rule: over-one, down-one. The explanation of this new-found competence would, in pre-Piagetian days, have been attributed to a black box called spatial ability, some children having more of it than others. Indeed, but for the grace of God, the diagonal could have appeared on any IQ test. In the paradigm I am endorsing, children's performance is to be explained in terms of representations and rules, that is, the beliefs, desires, and intentions implicated in their plans and goals.

The cognitive sciences are more or less defined by such analysis of the performances and judgments of intentional agents. Sternberg's (1977, 1984) studies of analogy, Case's (1992) studies of conservation, and Siegler and Robinson's (1982) studies of mental arithmetic are paradigmatic examples. Intelligence is not to be used to explain but rather is itself the thing to be explained. How, then, are we to explain skilled performance in any domain?

The answer, as previously suggested, is to be spelled out in terms of beliefs, desires, and intentions of conscious agents. Talk of intelligence, like talk of language ability, is to be "cashed out" in a mature science in terms of the processes and procedures, the concepts and rules applied in the domains under discussion. This is no less true for more general domains such as talking, problem solving, and acting than for those domains shaped by technological inventions such as writing and computing.

TECHNOLOGY AND INTELLIGENCE

Technology is explicitness, as Giedion (1948), whose history of mechanization remains unsurpassed, once argued. He established this claim by tracing the historical shift from "handicrafts" to mechanical production in the 18th and 19th centuries. What had remained implicit in an imitated skill had to be spelled out in terms of a set of explicit procedures that could then be mechanically applied. What had previously been carried out by skilled craftsmen and passed on through apprenticeships, came to be analyzed explicitly in terms of a set of rules and embodied in such technologies as the Jacquard Loom.

The cryptic suggestion that "technology is explicitness" is equally valid for one of the most studied of all cognitive technologies—namely, writing. Again the question is not does technology make us think better or make us more "intelligent" in the old sense. Consequently, the old question of whether modern man is more intelligent than his less civilized forebears is misleading. Rather the question is one of how the new technology transforms or otherwise relates to the existing actions and practices; that is, how does writing and learning to read and write relate to one's existing speech competencies? Do these transformed competencies allow one to do old things in new, more successful ways? And to what extent do they bring new goals into view? In these ways, how does writing make us more intelligent?

To even address such a set of questions it may be useful to distinguish a technology such as writing from such natural systems as language. Early work on technology was muddled by trying to defend the view that speech was a technology in that it was a means of operating on preverbalized thinking. Rather, I would restrict the concept of technology to invented artifacts and techniques that are explicit forms of traditional social practices. Thus, we may examine writing as an explicit representation of the implicit and largely unconscious practices involved in speaking. Similarly, we may examine notations for numbers as an explicit rule-based representation of the social practice of quantifying.

The historical development of a written notational system for numbers has been advanced by Damerow (1996), who argued that cognitive change should be seen as a move from more specific and local rules to more general, unified rule systems. By examining the notations for numbers of ancient Mesopotamian clay tablets, he was able to reconstruct the shift from the variable systems used to count different categories of objects to the development of a general system of numbers and computations applicable to all countables. He proposed that the development of an invariant system of counting and computing resulted from the need to improve bookkeeping operations in the ancient Babylonian temple economy. Babylonian mathematics developed when these general rules and operations became dissociated from their uses in accounting. Olson (1994) argued that the concept of zero was a byproduct of these notational practices.

The social practices in which writing is involved are both personal and social and include not only using writing to organize and clarify one's thoughts, but also for sending messages across space and through time and, perhaps most important, for collecting, storing, and organizing the documents that regulate large-scale social practices such as law, government, and the economy. These documentary practices have evolved over a long period of historical time, and they are spread through the society by systematic instruction in the school (Olson, 2003).

Clearly, many of the activities that we would consider to be intelligent, such as organizing an argument into strict logical form or explicitly defining the terms of one's discourse, are tied to our literacy practices. The specialized uses of mind that modern societies recognize as exhibiting intelligence can be seen in two levels of our literate activities: those pertaining to learning to read and write and those involved in using written documents to shape complex social practices.

LITERACY AND INTELLIGENCE

In learning to read and write one is learning not only a skill but learning to think about language and mind in a new way. This learning is summarized in the concept of metarepresentation. Whereas language is about, and in that sense represents, the world, writing is a representation of language, hence, a metarepresentation (Adams, Freiman, & Pressley, 1998; Homer & Olson, 1999). To oversimplify somewhat, one is learning to think not only about the world but learning to think about one's representations of the world. Literacy is, in this sense, a protypically metarepresentational way of thinking. What, precisely, do speakers know when they know a language? Chomsky (1980) hedges the claim that speakers know a language by offering a specialist term—*cognize*—to characterize this knowledge. It is a kind of knowledge in practice or implicit knowledge that one has in being a speaker of a language. I did not know that I routinely say "Where's my glasses" until

I was told that I should say "Where *are* my glasses?" To cognize English is to have some mastery of the grammar and phonology of English. Know in this implicit or tacit sense is a knowledge of the conventions of use that allow one to know what to expect in the performances of others and to know how to generate utterances that meet the expectations of others (Lewis, 1969). Not all of one's speaking practices remain implicit. The knowledge of English includes some metalinguistic concepts for referring to speech and language as indicated in such speech act verbs as *ask, say, tell,* and their cognates. But much of our knowledge of English remains implicit, remaining to be explicated in the process of learning to read and write.

Becoming literate requires that one begin to summarize across this implicit knowledge in such a way as to form a new class of entities, to summarize across the /b/ sounds in *baby, bobby, bib, balloon* to form a class of sounds to be represented in print by the letter *b*. It is, understandably, easier to form that class and learn the letter representing that class if one already has that phoneme in one's language competence, and it is easier to learn that classification if one is invited to do so by the presence of a sign in one's writing system denoting that class. Although there is considerable debate as to whether the analysis of speech into the categories of sound represented by the written signs is a prerequisite to learning to read (a view favored by phonics-based reading programs) or whether that knowledge is a consequence of learning the visual signs (a view favored by meaning-based reading programs), there is little dispute that reading involves a new kind of metalinguistic awareness, namely, an awareness and classification of the phonological properties of speech into the categories provided by the script. Such phonological awareness is useful for reading and spelling but has little relevance to thinking more generally (unless filing things alphabetically is seen as an intellectual process). An awareness of other, higher-order features of language—words and sentences and the documentary practices premised on them such as making lexicons, dictionaries, logics and mathematics—has broad conceptual uses and in that sense contributes to and alters intelligence. More precisely, these practices give intelligence its modern character.

Consider one's knowledge of words as conceptual objects. The idea that the stream of speech could be represented in terms of a limited inventory of words is an idea that is a byproduct of literacy. It is an idea that makes the concept of a dictionary possible, and dictionaries are, transparently, devices with which to think. (Need I point out that one performs much better on the vocabulary section of an IQ test if one has access to a dictionary?) Studies by Ferreiro and Teberosky (1982), Vernon and Ferreiro (1999), and Homer and Olson (1999) showed that prereading children take written signs to represent objects and events rather than to represent linguistic units, namely, words. Prereading children do not know what a word is or that an utterance is comprised of a string of words until they learn a script. If one reads an illustrated story to a prereading child containing an expression such as "my porridge is too hot," and then asks the child where it says that, the child is more likely to point to the picture of the hot porridge than to the printed text itself. Similarly, if asked "Can you write 'no cats'?" a prereading child is likely to say, "You can't write that because there are no cats." That is, the child is aware of the object of discourse, the cat or the porridge, but not of the linguistic utterance expressing it (Bialystok, 1986; Olson, 1994). Scribner and Cole (1981) found that whereas readers of a syllabic script could segregate the stream of speech into syllabic units, they could not do so into words. In one sense, of course, speakers "know" the words of their language in that they are units governed by the syntactic structure of sentences. However, the ability to segment utterances into words and to think about words as conceptual objects is largely, if not exclusively, tied to knowledge of writing a word-based script. Saenger (1997) argued that the interposition of spaces between words, something that was added to written Latin in the late Middle Ages, not only made silent reading possible, but also enhanced reflection on words and their meanings.

Some developmental research also bears on the issue of consciousness of words. Doherty and Perner (1998) found that whereas 3- and 4-year-old children could readily agree that the synonyms *rabbit* and *bunny* both may refer to the same entity and that the hyponyms *rabbit* and *animal* could similarly be used to refer to the same object, they were unable to operate on the representations themselves. Thus, after the experimenter named the picture with one of the words of the pair, children were given the task of naming the picture with the other word by being instructed to "say something different." Three- and four-year olds were unable to do so. By 5 or 6 they could not only do this task, but could also answer such questions as "What else could this be called?" or "What's another word for this?" and the like. That is, children in such a literate environment learn not only to think about words but about the relations among words—synonyms, antonyms, hyponyms—that had previously remained implicit. The extent to which this skill can develop in a nonliterate environment remains to be explored.

Further, it may be argued that an explicit knowledge of words and their semantic relations is what makes possible the concept of literal meaning, the meaning that is bound, so far as is possible, to the linguistic form of an expression. The interplay between literal meaning and intended meaning, between what is said and what is meant (Olson, 1994), gives rise to the kind of prose taken as paradigmatic in bureaucratic societies, the language of constitutions, charters, contacts, and law as well as of science. Such language is not an immediate and necessary implication of writing but rather a specialized use of writing for formulating documents and other records that serve an archival function. Such archival texts evolved over a long period of time in a particular cultural setting (Johns, 1998). The extent to which such language is universalizable is of much interest currently as treaties and agreements are negotiated between radically different cultural groups. Treaties and contracts are necessarily precise and, so far as possible, not open to interpretation. The genre of discourse that evolved to meet this requirement was dependent not only on the availability of writing but on a particular orientation to written language. Further, such documents depend on the invention or development of more formal, bureaucratic institutions for regulating the uses of those documents.

Bureaucratic institutions are necessarily literate ones. Sociologists, following Tönnies (1887), distinguish between small-scale and large-scale societies. Small-scale societies are community-based (Gemeinschaft) societies in which relations are between persons known to each other and rely on trust, goodwill, and, ultimately personal authority. Large-scale societies, on the other hand, are impersonal, bureaucratic organizations (Gesellschaft societies) that are based on roles, rules, and credentials that are designed to coordinate the activities of strangers across space and time. Such institutions are literate institutions in that they rely on the creation and interpretation of important documents that spell out the roles and rules and the lines of authority and responsibility that are necessary for the functioning of a complex organization. To function intelligently in such an organization is to base one's decisions on the rules and principles adopted by that organization. The wise man gives pride of place to the bureaucratic functionary, the one who knows and follows the rules that allow the decisions to be spread or distributed across the organization, including the policy branch, the records and information branch, the finance office, and the like. In a modern university, for example, each discipline is responsible for a specific range of concerns; no one person is responsible for everything. The discipline, in turn, is comprised of credentialed professionals attached to the appropriate archive of records and materials, and so on. The ability to perform in any of those professional roles depends on the competence to create and interpret documents of the approved type in that institutional context. Thus, there comes to be a variety of literacies, the specialized competencies needed to cope with, for example, scientific, legal, or economic documents. Modern societies succeed just by such division of labor and specialization of function, and persons succeed just by being able to perform specialized roles in such a scheme.

It is in these bureaucratic contexts that the competencies we now characterize as intelligent have emerged. The ability to define a word is a basic skill only in relation to the competencies relevant to the social practices of a document-based society. Intelligence tests are achievement tests that examine the extent to which one can perform the literate activities required in a bureaucratic society. This may be seen by returning to the relation between literacy and judgments of synonymy, antonymy, and hyponomy discussed earlier. Knowledge of the relation between words is a central feature of most IQ tests. Correct answers require metarepresentational competence, knowledge of the definitions of words, relations between words, as well as strict logical entailments, abstract representations of spatial relations, and the like. To score well on the vocabulary items of a typical intelligence test (e.g., Stanford-Binet) one must give not only a description—A bicycle has wheels—but also provide a superordinate hyponym—A bicycle is a thing to ride on—or A bicycle is a vehicle. Watson and Olson (1987) observed children being taught just this superordinate function in a classroom context, as follows:

TEACHER: What's a lullaby?
CHILD 1: It puts you to sleep.
TEACHER: But what *is* it?
CHILD 2: It's a song.
TEACHER: Right.

Note how the teacher rejects the first child's expression even if it is correct. The teacher repeats the question emphasizing the existential verb and only when the second child provides the hyponym does she accept the answer. This is the answer to which an IQ test scoring manual would give full marks. It is not difficult to see why IQ test items predict school success and vice versa; schools teach the very structures that IQ tests test. Intelligence is, therefore, an achievement.

Intelligence test items are not culturally neutral items as a whole generation of cultural psychologists has clearly shown (Cole, 1996). Indeed, it is reasonable to suggest that there is no such thing as a culture-fair test of intelligence. Rather the items that find their way into intelligence tests are those that serve as indications of those competencies relevant to participating in a modern literate society. This is not to deny that such tests may be useful, especially if some more specific assessment is not available. They have predictive value in a literate society because the competencies sampled bear a direct relation, first, to the advanced studies required for credentialling and, then, to the actual literate activities involved in participating in the bureaucratic institutions of a modern society. Nor is it to deny that such competencies are rooted in our biology. Population geneticists' studies of heritability show convincingly that variability exists in whole populations that may be traced to variability in the genetic code. Rather it is to argue that in our attempt to understand why a person's speech or actions take the form that they do, the answer is to be found by examining their beliefs, understandings, and the rules and procedures they have been taught in school and in the roles they learn to play in the society at large.

On one hand, the competence modern societies characterize as intelligence is little more than the possession of the specialized knowledge associated with literacy; intelligence is "skill in a medium of representation" (Olson, 1970, p. 193; 1986). Yet, on the other hand, writing is a defining technology of a modern bureaucratic society and competence with that technology and its uses constitutes a critical part of intelligence. By attending to the cognitive aspects of literacy we may account for not only some of the features of the modern mind but also why measures of intelligence take the form that they do in such a society.

REFERENCES

Adams, M., Freiman, R., & Pressley, M. (1998). Reading, writing and literacy. In W. Damon (Ed.), *Handbook of child psychology*, Vol. 4, Series edition. I. Seigel & K. Renninger (Eds.), *Child psychology in practice*, 5th ed. (pp. 275–355). New York: Wiley.

Bialystok, E. (1986). Children's concept of word. *Journal of Psycholinguistic Research, 15*, 498–510.

Bruner, J. S. (1960). *The process of education.* Cambridge, MA: Harvard University Press.

Case, R. (1992). *The mind's staircase.* Hillsdale, NJ: Lawrence Erlbaum Associates.

Chomsky, N. (1980). *Rules and representations.* New York: Columbia University Press.

Cloud, J. (2001, March 12). Should SATs matter? *Time, 157*, 62–70.

Cole, M. (1996). *Cultural psychology.* Cambridge, MA: Harvard University Press.

Coyne, J. (2000, April 27). Not an inkling: Review of Matt Ridley's "Genome." *London Review of Books, 22*, 13–14.

Damerow, P. (1996). *Abstraction and representation: Essays on the cultural evolution of thinking.* Boston: Kluwer.

Dewey, J. (1972). The reflex arc concept in psychology. In J. A. Boydston (Ed.), *John Dewey: The early works. Vol. 5:* 1895–1898 (pp. 96–109). Carbondale: Southern Illinois University Press.

Doherty, M., & Perner, J. (1988). Metalinguistic awareness and theory of mind: Just two words for the same thing? *Cognitive Development, 13*, 279–305.

Easton, M. (2001, February 12). Prostate cancer gene found. *University of Toronto Bulletin*, p. 3.

Ferreiro, E., & Teberosky, A. (1982). *Literacy before schooling.* Exeter, NH: Heinemann.

Gardner, H. (2001). Jerome Bruner as educator. In D. Bakhurst & S. Shanker (Eds.), *Jerome Bruner* (pp. 127–129). New York: Sage.

Giedion, S. (1948). *Mechanization takes command.* New York: Oxford University Press.

Homer, B., & Olson, D. (1999). The role of literacy in children's concept of word. *Written language and literacy, 2*, 113–137.

Johns, A. (1998). *The nature of the book: Print and knowledge in the making.* Chicago: University of Chicago Press.

Lewis, D. K. (1969). *Convention: A philosophical study.* Cambridge, MA: Harvard University Press.

Lewontin, R. (1976). The analysis of variance and the analysis of causes. In N. Block & G. Dworkin (Eds.), *The IQ controversy: Critical readings* (pp. 179–193). New York: Pantheon.

Olson, D. R. (1970). *Cognitive development: The child's acquisition of the concept of diagonality.* New York: Academic Press.

Olson, D. R. (1986). Intelligence and literacy: The relationships between intelligence and the technologies of representation and communication. In R. J. Sternberg & R. K. Wagner (Eds.), *Practical intelligence: Nature and origins of competence in the everyday world* (pp. 338–360). Cambridge, UK: Cambridge University Press.

Olson, D. R. (1994). *The world on paper: The conceptual and cognitive implications of writing and reading.* Cambridge, UK: Cambidge University Press.

Olson, D. R. (2003). *Psychological theory and educational reform: How school remakes mind and society.* Cambridge, UK: Cambridge University Press.

Piaget, J. (1950). *The psychology of intelligence.* London: Routledge & Kegan Paul.

Saenger, P. (1997). *Space between words:The origins of silent reading.* Stanford, CA: Stanford University Press.

Scribner, S., & Cole, M. (1981). *The psychology of literacy.* Cambridge, MA: Harvard University Press.

Siegler, R. (1992). The other Alfred Binet. *Developmental Psychology, 28,* 179–190.

Siegler, R. S., & Robinson, M. (1982). The development of numerical understanding. In H. Reese & L. Lipsitt (Eds.), *Advances in child development and behavior* (pp. 241–312). New York: Academic Press.

Sternberg, R. J. (1977). *Intelligence, information processing, and analogical reasoning: The componential analysis of human abilities.* Hillsdale, NJ: Lawrence Erlbaum Associates.

Sternberg, R. J. (1984). Mechanisms of cognitive development: A componential approach. In R. J. Sternberg (Ed.), *Mechanisms of cognitive development* (pp. 163–186). San Francisco: Freeman.

Tönnies, F. (1887). *Community and society* (C. P. Loomis, Trans. and Ed.). East Lansing: Michigan State University Press. (Original work published 1887)

Watson, R., & Olson, D. R. (1987). From meaning to definition: A literate bias on the structure of meaning. In R. Horowitz & S. J. Samuels (Eds.), *Comprehending oral and written language* (pp. 329–353). San Diego, CA: Academic Press.

Vernon, S., & Ferreiro, E. (1999).Writing development: A neglected variable in the consideration of phonological awareness. *Harvard Educational Review, 69,* 395–415.

II

COGNITIVE CONSEQUENCES OF EDUCATIONAL TECHNOLOGIES

4

Do Technologies Make Us Smarter? Intellectual Amplification *With*, *Of*, and *Through* Technology

Gavriel Salomon
University of Haifa, Israel

David Perkins
Harvard University

The impulse to make what you do not have runs deep in the human mind. Children design implements such as cranes made of sticks, string, and house keys, and transform pairs of socks into balls to play with. Such children's games are but a small sample of a vigorous human enterprise. From the dawn of civilization, people have created physical and symbolic devices that help them do what they cannot accomplish through bare flesh and bone: tools, instruments, machines, writing systems, mathematics, and on and on. Such products of human invention extend both our physical and our intellectual reach.

This much is not news at all. But out of such ordinary observations one can fashion a provocative question: Does technology make people smarter? More formally, do technologies expand our cognitive capabilities in any fundamental sense? To be sure, with the help of certain technologies we can see farther—optical and radio telescopes—and see smaller—optical and electron microscopes—as well as access the knowledge of the past and knowledge from the other side of the world with great convenience—libraries, the Internet. But it hardly seems reasonable that these should count as making us "smarter." Indeed, the comparison with the physical assists provided by some technologies is discouraging. We do not ordinarily count ourselves transformed from the proverbial 97-pound weakling to Charles Atlas simply by sitting in the cab of a bulldozer. Why should sitting at a computer terminal score any differently?

Still, from another perspective, the question should not seem too bold or bizarre. After all, some cultural artifacts have been argued, even shown, to affect minds. Thus, for example, literacy has been claimed to modify minds by teaching abstract thinking (e.g., Greenfield, 1972); literacy is also said to facilitate the development of hermeneutics—the distinction between what is said in a text and what is interpreted on its basis (Olson, 1986); and large-scale processes of modernization of the kind studied by Alexander Luria in Central Asia are claimed to account for the development of abstract thinking (Luria, 1976). Some technologies offer new metaphors to think with—"the brain as a computer" (Bolter, 1984), whereas statistical tools are said to lead to the development of psychological theories (Gigerenzer, 1991). Given such claims and observations, would it not make sense to ask whether technologies, perhaps some technologies under some social and psychological conditions, may affect the intellectual capabilities of some minds in some relatively lasting ways?

Gaining encouragement from such examples, this chapter examines whether and in what senses technologies might make us cognitively more capable. Naturally, approaching such a question in a reasonable way requires staking out the territory: What kinds of technologies do we have in mind? Technologies make us cognitively more capable in what senses? After explaining the particular perspective adopted here, the discussion builds on our previous work by offering a three-way framework to address the question. Considered are *effects with* technology, how use of a technology often enhances intellectual performance; *effects of* technology, how using a technology may leave cognitive residues that enhance performance even without the technology, and *effects through* technology, how technology sometimes does not just enhance performance but fundamentally reorganizes it. We compare and contrast these three modes, pondering their typical timelines, their frequency of occurrence, the magnitudes of their impact, and related points. Finally, the analysis turns to a particularly provocative case previously mentioned: how technologies offer new metaphors to think with. The conclusion positions this kind of development within the framework of *effects with, of,* and *through,* and concludes with a broad assessment of the senses in which certain technologies truly may be said to make us smarter.

FRAMING THE PROBLEM

As just noted, one can hardly address such a question without some clarification of the question itself. First, then, what sorts of technologies are our focus here? There are many candidates for this concept, such as technical tools (e.g., the pencil); symbol systems (e.g., the spoken language, the language of film); the sciences and their notations (e.g., mathematics); and "intelligent" (or partly intelligent) instruments (interactive concept mapping tools).

Too broad a focus is to be shunned as affording too glib an answer about technology making us smarter. After all, many technologies might be said in one way or another to enhance cognitive functioning—for instance, medical technologies or nutritional technologies, which improve cognitive functioning as a side effect of improving general health. However, such indirect effects fall far from the present focus. Also set to one side are technologies that just put things closer in space and time—like telescopes, the printing press, and the telephone—though they, too, have cognitive impacts through making information of diverse sorts more readily available. The emphasis in this analysis falls on technologies that directly facilitate or even carry out cognitive work—calculators, statistical packages, word processors, outliners, and the like—as well as symbol systems with which one can think—writing, mathematical notation, musical notation, and so on. In the course of the analysis, this rough staking out of the territory will become more specific.

Furthermore, one can hardly ask whether technologies make people smarter without clarifying what "smarter" amounts to. We certainly do not mean to examine simply whether certain technologies raise people's IQ. Not only is IQ a highly controversial construct, but it is also one among various constructs used to express an essentialist position on what it is to be smart. That is, truly being smart is something not to be identified even in part with something like having a good flexible repertoire of cognitive strategies but rather with something deep in the fundamental mechanisms of cognition—say, highly efficient neural processing.

We, along with many others, have argued that attempts to reduce everyday intelligence or thoughtfulness or acuity to some essential mechanism fail to make a full and convincing case (Grotzer & Perkins, 2000; Perkins & Ritchhart, 2004). Accordingly, we adopt a performance view: The fundamental question is not "How fast do your neurons work?" or something in that spirit, but rather what kind of cognitive performance do you display—"How well do you solve problems and make decisions?" or "How quickly and with what sensitivity do you perceive complex environments?" More capable cognition might reflect a combination of rather different cognitive resources, such as problem-solving heuristics, helpful conceptual systems, metacognitive self-management, rich and flexible perceptual categories, and, to be sure, any core cognitive mechanisms one wants to toss into the mental pot. Such skills and abilities should show some reasonable range of generality to count as part of being smarter, although they need not be nearly as general as *g*. Also important to include here is the dispositional side of cognitive functioning: Good thinkers are attentive, persistent, alert to needs and opportunities, and so on. Evidence suggests that disposition is just as important to what it is to function in a smart way in the world as are various abilities (Perkins & Ritchhart, 2004; Perkins, Tishman, Ritchhart, Donis, & Andrade, 2000; Ritchhart, 2002).

With the issue framed in this manner—technologies that have the potential of more or less directly facilitating cognitive work and a performance conception of cognitive capability—the question of whether technology makes us smarter becomes approachable.

EFFECTS *WITH* TECHNOLOGY

How then might technology affect the intellect? Consider the case of computers. There is surely difference among improving writing performance with the use of a word processor, coming to search for information in entirely new ways, or learning from a dynamic model builder how to think in new ways that reflect the "thinking" of the tool (Cline & Mandinach, 1994; Salomon, 1979/1994). Such possibilities as these are the focus of our interest here as they touch on the wider question of technology and mind: Does technology shape minds? In earlier work, we (Salomon, Perkins, & Globerson, 1991) distinguished between two ways in which technology affects minds: *effects with* technology, manifested by amplified performance while one is operating a tool, and *effects of* technology, manifested by changed skill mastery that comes as a consequence of that activity with the tool, even without the tool in hand. Let us examine *effects with* technology and how this concept bears on the central question.

Effects with technology emerge through the interaction when certain intellectual functions are downloaded onto the technology (spelling, computing, ready rearranging), thus establishing an intellectual partnership with the user (e.g., Pea, 1993). By partnership we imply a division of labor and an interdependence typical of the interaction with tools (e.g., automobiles, databases), which we have to skillfully operate, as contrasted with machines (e.g., watches, refrigerators) that usually work for us without too much involvement on our part (Ellul, 1964). The partnership becomes intellectual to the extent that cognitive functions—such as computing, mapping, integrating or composing—are distributed between the tool and the individual using it (e.g., Perkins, 1993). To the extent that such a partnership frees the user from the distractions of lower-level cognitive functions or ones that simply exceed mental capacity, and provided that the tool is used in mindful ways that benefit from the partnership, it is likely to lead to improved intellectual performance.

A case in point is the Norwegian computer-enriched approach of *Writing for Reading* (Trageton, 2001) whereby 5-year-olds learn to write on the computer long before they learn to read. The quality of their essays while writing with the computer far exceeds that of their peers in more traditional literacy classes. Another case in point concerns the search activities on the Internet adopted by college students (Cothey, 2002). Turning from students to professionals, contemporary technologies afford endless examples. Spreadsheets allow creating dynamic financial models that permit exploring alternative scenarios with a fluency and flexibility impossible to match by hand. Symbolic computational systems like *Mathematica* foster the ready generation of mathematical derivations and close inspection of the behavior of functions, again in a ways hard to manage by hand. Concept mapping software allows constructing complex webs of relationships that would be exceedingly difficult to envision mentally or represent through conventional sentences and paragraphs.

However, we do not need to look at computational technologies to find examples. Recall that, in our framework, symbol systems count as technologies. Way before computers, the development of mathematical notations of various kinds enabled lines of mathematical inquiry that otherwise would have faltered for lack of a vehicle. Text itself, besides providing a channel of communication, also has long functioned as a vehicle of thought, as, for instance, people laid out arguments on topics from anthropology to zoology, assessed them, and improved them. The sketches of an architect or engineer enable exploratory processes that would be impossible through mental imagery and premature for actual constructed prototypes. Although certainly computers have provided powerful new resources in support of thinking, most of those resources are presaged by paper-and-pencil symbol systems that already gave thinking a substantial boost.

So, yes, working with certain technologies makes us smarter, at least in the sense that it leads to smarter performance. Indeed, at this point one can characterize a little more precisely the kinds of technologies that serve this role: They are what might be called cognitive technologies, technologies that enhance cognitive functioning through directly affording cognitive support rather than as a side effect through, say, enhanced health.

Perhaps the most natural rejoinder to this position is, "But people are not really any smarter just because they are using a spreadsheet or Mathematica in a reasonably skilled way." To be sure, people have not necessarily acquired any general cognitive capabilities in the absence of the technology (however, see the discussion Effects of Technology in the following). However, the "not really" also betrays an inclination toward an essentialist conception of being smart as if nothing counts as smarter but the bare brain functioning better. Yet, the success of human beings in this world plainly does not depend on bare brains any more than it depends on bare hands. It is the dramatic flexibility of the brain and the hand to fashion tools and use them in so many varied and powerful ways that is perhaps the most distinguishing mark of the human condition. The average human being does not function as a person solo but overwhelmingly as a "person plus"—plus physical and symbolic support systems, and also plus a web of social relationships, although that is not focused on here (Perkins, 1993). Complex human cognition is typically distributed cognition—distributed over social and physical support systems (e.g., Hutchins, 1995; Salomon, 1993a, 1993b). Person plus is the norm for the human condition, and human beings as intellectual agents are best considered not stripped of, but suitably equipped with, tools.

Another reservation might point to usages of cognitive technologies that seem not to lift cognitive functioning at all. Writing quick, newsy letters or searching on the Internet for movie reviews may prove convenient but would not seem in any dramatic way to enhance cognitive functioning. Musical and mathematical notations as a means simply of publishing symphonies and proofs amounts to enhanced communication but not amplified cognition, in contrast with those same notations as symbolic scaffolds for individual and group inquiry and expression. In other words, if cognitive technologies support cognition sometimes, they certainly do not always do so.

The best response to this is to note that, just as the bare brain is not quite the right unit of analysis, neither is the bare technology. When speaking of technology and the intellect, we address not so much the technology itself but, as Ellul emphasized in his classic *The Technological Society* (1964), various skilled uses of the technology in interaction with it. What makes for a cognitive technology in the sense outlined is not the technology alone but cognitively demanding systems of activity it typically enables. We follow here the conception offered by Scribner and Cole (1981) according to which technologies are seen as part of systems of particular activities. Thus, one cannot speak of word processors independently of their use, say, for transcribing dictation or communicating casual messages (not enhancing complex thought) versus constructing complex arguments. Similarly, it would be strange to speak of the World Wide Web, let alone its intellectual consequences, without addressing with some differentiation the diverse activities it supports. After all, it is the activities with a technology that might affect the intellect, not the technology per se. As Scribner and Cole (1981) pointed out when discussing the possible intellectual consequences of literacy, "The nature of these practices, including of course their technological aspects, will determine the kinds of skills ('consequences') associated with literacy" (p. 236).

In summary, part of the answer to the question—does technology make us smarter?—boils down to this: Cognitive technologies—technologies that afford substantial support of complex cognitive processing—make people smarter in the sense of enabling them to perform smarter. Moreover, given that human beings are by nature toolmakers and tool users, this is a pretty reasonable sense of smarter. That said, it is also important to ask whether experiences with cognitive technologies can develop cognitive capabilities that remain available without the tool at hand. This brings us to the complementary theme of *effects of* technology.

EFFECTS *OF* TECHNOLOGY

Effects of technology, as you will recall, concern effects, positive or negative, that persist without the technology in hand, after a period of using it. For example, one might ask whether there is an improvement in general writing or reading ability in the *Writing to Read* example, or a tendency to be more (or less) systematic in information search in general as a consequence of searching the Internet.

If we ask about technologies in general, there is no systematic trend toward positive or negative *effects of*. The blacksmith may develop more brawn by wielding his tools, but the bulldozer driver will not; and the suburbanite who drives everywhere may grow weaker because of his powerful car. Using refrigerators and ovens does not sharpen our thermodynamic reasoning much, nor does air travel teach us aerodynamics. However, the present analysis focuses on technologies used in a tool-like way, unlike refrigerators and airplanes, and on cognitive technologies, specifically, as characterized earlier, technologies that form a cognitive partnership with the user. So in this case, the prospects would seem to be brighter.

One would look for *effects of* as a consequence of interacting with a technology—the acquisition of a new skill (or becoming de-skilled in some way) or the improved mastery of an existing one. A subcategory of such effects would be the acquisition of specific technology—or tool-related skills (e.g., leaning to navigate the Internet). However, we are less concerned here with such specific effects than with the possibility of developing more generalizable skills that, while cultivated by the interaction with the technology, become sufficiently general to allow applications that transcend the technology-related context. A candidate case might be the cognitive effects of programming anticipated by researchers in the 1980s, a case about which we will comment later.

Then, from an empirical perspective, what signs are there of *effects of*? Salomon (1979/1994) carried out a series of experiments and field studies to test the hypothesis that active exposure to the unique symbol systems elements of film and television can become internalized to serve as more generalized cognitive modes of representation and operation. These studies were based in part on the rationale advanced by Bruner (1966) according to which:

- Man is seen to grow by the processes of internalizing the ways of acting, imagining, and symbolizing that "exist" in his culture, ways that amplify his powers (p. 320).
- Any implement system, to be effective, must produce an appropriate internal counterpart, an appropriate skill necessary for organizing sensorimotor acts, for organizing percepts, and for organizing our thoughts in a way that matches them to the requirements of implement systems. These internal skills, represented genetically as capacities, are slowly selected in evolution (p. 56).

Thus, for example, in one experiment Salomon (1979/1994) showed that school-age children manifest significantly improved ability to interrelate perceptual parts and wholes as a consequence of guided exposure to filmic zoom-ins and outs. In another study, Salomon showed that children significantly improved their ability to change visual perspectives as a result of exposure to angle-changing camera movements. Salomon concluded from these and similar studies that children can and do appear to internalize symbolic forms from the visual media and use these as cognitive tools.

To turn to other cases, some researchers and educators in the 1980s seriously explored how mastering the programming of computers might enhance thinking. The notion was that the cognitively complex and challenging activity of programming provided a kind of mental gymnasium, both exercising and drawing students' attention to patterns of analytical and diagnostic reasoning. Research on such interventions generally proved discouraging: No significant impact was found. However, in a few cases gains on transfer tasks did appear. The pattern of contrast between the negative and positive cases was revealing: The positive cases included not just the programming experience but also features that encouraged reflective abstraction, along with sufficient length and depth of experience with programming to develop a reasonable skill set. We caution here that the occasional positive findings do not recommend conventional programming experience as a particularly powerful or efficient approach to developing thinking skills. Many other more direct approaches have a much better track record (Grotzer & Perkins, 2000; Ritchhart & Perkins, in press). However, this research does offer clear instances of *effects of.*

Another recent finding offers a clear instance of effects of technology use on complex cognition. Researchers of avid players of video games examined male students, ages 18 to 23, comparing those who played action-oriented video games at least 1 hour a day, 4 days a week, with those who rarely played (Green & Bavelier, 2003). The video game enthusiasts proved to be greatly superior at a range of tasks that involve rapid visual processes, such as finding a target object in a messy scene. To check whether these results truly represented experience with the video games, in contrast with the game environment attracting people with such skills, the researchers conducted another study where they trained both men and women to play action video games for 10 days, 1 hour per day. The trainees showed substantial improvement on the perceptual tasks, without reaching the levels of performance displayed by the avid gamers. Although the perceptual tasks were plainly cognitively challenging, the researchers noted that they did not address deliberative thinking.

Formal research aside, some *effects of* are commonplace. Recalling that we consider notational systems technologies, familiarity with music notation does more than enable analysis and composition; it also shifts to some extent how one hears music, even without pencil and score in hand. Indeed, there are cognitive technologies that are meant to be withdrawn—rather like the training wheels sometimes used for learning to ride a bicycle. For instance, children learning to write in Hebrew normally begin with an alphabet that includes extra marks for vowels, marks that are later withdrawn. In the same spirit, formal grammatical rules often assist second language learners in initial mastery of the language and enable reasonably accurate performance early on; but, with practice, the rules in their explicit form fade away and may even prove difficult to recall, as the learner advances to automatized fluency. In such cases as these, the supportive technology—the notations, the rules, the training wheels—is designed for temporary *effects with* leading to lasting *effects of.*

EFFECTS *THROUGH* TECHNOLOGY

The two categories of technology's effects—*effects with* and *effects of*—as originally outlined by Salomon, Perkins, & Globerson (1991), seemed a sufficient account at that time. However, further reflection by us and by others suggests that another fundamental distinction deserves attention. To be sure, *effects with* technology enhance the performance of the activity in question, and such gains are most welcome. However, from time to time the impact of a new technology is more radically transformative. Consider, for instance, the reorganizing impact on warfare of wave after wave of technological advance, from the longbow to the crossbow to the rifle to airplanes and tanks to nuclear weapons. A rifle is not just a better longbow. Over and over again, technical innovations have led to fundamental restructuring of how battles are fought. Or consider the impact of concrete on Roman construction—and construction in the modern era—an innovation that enabled kinds of structures and processes of construction thought to be unimaginable.

In general, the use of new technologies qualitatively and sometimes quite profoundly reshapes activity systems rather than just augmenting them. This we name *effects through* the use of technology. Turning to cognitive technologies, one prime example comes from the long tradition of scholarship in the area of literacy, showing that reading and writing "reorganized the process whereby we retrieved, compared, listed, and ordered our ideas and, eventually, transmitted them to you" (Cole & Griffin, 1980, p. 363).

Cole and Griffin (1980) argued that the concept of amplification, as formulated by Bruner (1966), implies a changed intensity of an action but not any qualitative change. Moreover, they distinguished between improved *performance* as a criterion, whereby a child with a pencil would show "improved memory," and the *process* through which better performance is produced. The child with the pencil does not have any improved memory capacities by himself or herself, but the task of holding on to information has been qualitatively restructured. Third, testing for any cognitive residue (*effect of* technology) would require that one tries to operate without the technology: Write without a pencil, organize information without tools that allow tabulation, or compute without a calculator. In some cases this is possible, but in many others the very execution of the activity presupposes the existence of that tool that enables the activity. How can you compare the improved "net" killing ability of rifles relative to bows and arrows in the absence of either one of them? Thus, Cole and Griffin wrote, "Central to the present argument... it is unnecessary to posit a general change in internal cognitive activity as a consequence of literacy—the effect requires that the tool be in the user's hand" (1980, p. 358).

Extending this idea to modern computational technologies, it is not difficult to identify a number of ways in which they do not simply enhance but reorganize performance. Consider, for example, scientific inquiry. Classically, physics and related disciplines model phenomena through mathematical equations, but computational technologies allow a new kind of theory: the rule set that guides a simulation, with predictions generated by running the simulation. One notable recent example of this is Wolfram's (2002) proposal that science can be reconstructed around cellular automata, a proposal greeted with considerable controversy but nonetheless illustrative of the point. Contemporary software for architectural design allows a fluency in exploring revisions and alternatives that changes how architects can relate to their multiple constituencies. Speaking of *effects through*, a client today can experience "walk-throughs" of proposed structures and interact in a concrete way with the architect as never before.

Hypermedia are expanding our conceptions of what it is to author. Although people certainly post conventional essays, stories, and poetry online—after all, these are robust, discursive, and expressive forms—many Web resources gain their power and flexibility from the ready linking allowed by the medium and invite Web authors into new realms of craft and imagination as they explore the affordances (Bolter, 1991). Teamwork mediated by the Internet enables geographically dispersed projects to proceed with extensive interaction and tight coordination, and only an occasional face-to-face contact. And so on.

The original scheme of *effects with* and *effects of* would have classified all such examples as *effects with.* The new distinction between *effects with* and *effects through* reflects the reorganizing impact on the activity system in question—whether for scientific theorizing, designing architecture, collaborating, or something else—wrought by some new technologies. That said, it should be noted that the contrast between *effects with* and *effects through* is not so much categorical as polar. Reorganization is a matter of degree. Sometimes a technology changes things a little, sometimes a little more, and sometimes a lot. However, the existence of a fuzzy border does not really trouble the present inquiry. Our mission here is to offer a broad perspective on the impact of cognitive technologies, and haggling about borderline cases is of less concern than recognizing the range from *with* to *through.*

Indeed, it is common to find the whole range expressed within current users of a single technology. For example, does writing with a word processor fundamentally change the act of authoring? The answer depends on how much of a change you count as fundamental, but also on who the writers are, how much time they have had to explore the affordances of the new technology, and how aggressively they have done so. Thus, students new to word processors tend to use them in rather routine ways, for spell checking, minor textual revisions, and neat printing (Daiute, 1985). None of this should be disdained: It can be very motivating and engaging. However, with experience and mentoring, students can come to use the affordances of the word processor to allow large-scale structural revisions, which they are not likely to undertake otherwise because it is so very inconvenient.

The long ramp up to *effects through* marks what Perkins (1985) referred to as the "fingertip effect"—the seductive assumption that simply making a technology available quickly draws users into a flexible use of its full affordances. On the contrary, the tendency is to assimilate new technologies into old patterns of practice, yielding a very modest version of *effects with.* Learners need time and guidance to achieve the effects that many contemporary cognitive technologies afford. From a longer-term historical perspective, the most dramatic *effects through* are not likely to be apparent at all during the early years of a new technology. It takes time for innovators to see the possibilities, time for early trials, time for a kind of Darwinian sifting of those new ways of working that truly offer a lot, and time for the new ways of working to pass into widespread use.

COMPARING EFFECTS *WITH, OF*, AND *THROUGH*

We have taken stock of *effects with, effects of,* and *effects through* cognitive technologies one by one. Each category represents a way in which cognitive technologies might be said to enhance people's cognitive capabilities—to "make us smarter." The three apart are like pieces of a puzzle, worth putting together to get the big picture. Comparing the three with one another, what relative magnitudes of impact can we anticipate and how quickly can we expect such effects to emerge?

Concerning pace, *effects with* is the clear winner. *Effects with* generally emerge fairly quickly, as one masters the rudiments of word processing, spreadsheets, hand-held calculators, and similar cognitive technologies. In comparison, *effects of* and *effects through* develop over longer periods of time. The relatively quick yield provided by *effects with* is to be expected. This is the classic consequence of tools: Put a rifle or a wheelbarrow in a person's hands, and almost at once the person becomes a person-plus, more capable by a quantum leap.

That said, it is also important not to overestimate the impact. Rifles may make people immediately more deadly in certain ways but do not make them good shots. Likewise, word processors and spreadsheets quickly create gains in capability, but certainly not expert performance. As with any challenging area, the ramp to expertise is a long one. Recalling the previous reference to the fingertip effect, it is naïve to suppose that simply providing the technology leads smoothly and quickly to a wide-ranging exercise of its full affordances. Moreover, even over time many individuals may not develop notable expertise. Like rifles or golf clubs or hammers, cognitive technologies can easily be used in casual ways for years, without any striking advance in sophistication. What technologies afford they do not typically demand. Perpetual duffer-hood is the fate of many a user of cognitive and other technologies.

Concerning relative magnitude of impact, *effects with* technologies are again to be prized for their immediate payoff as well as the further improvement that follows over time with serious and thoughtful use. However, *effects through*, virtually by definition, harvest the full transformative potential of cognitive and other technologies, as over longer periods of time individuals and groups explore the further reaches of their affordances in ways that lead to reorganized systems of activity.

The notable loser in this comparison of pace and magnitude is *effects of* technology. Such effects generally seem to be both of modest magnitude and slow to emerge. As argued previously, *effects with* technology generally overshadow greatly any *effects of*. Moreover, extended periods of usage seem necessary for *effects of* to accumulate and generalize. Indeed, in many cases it is hard to say whether there are substantial *effects of* at all, and in other cases it is tempting to suggest that, if there are, they are not worth much attention, as they fall short of *effects with* and *through*. This is a curious conclusion, because often in writings about the impact of technology, it is *effects of* that authors appear to have most in mind, although such effects do not appear to be the big win!

Accordingly, it's worth spending a moment to examine *effects of* further and understand more about their meager showing. Whereas there are many examples of *effects with* tools and *effects through* tools, there is a great paucity of studies and findings about *effects of* technology. There are at least two reasons for this lack. First, it is methodologically difficult to demonstrate the "net" effect of tool usage on the development of generalizable skills and tendencies. Hardly ever can the effects of tool use be studied in total isolation from other variables.

Second, even more challenging is the fact that technology's effects on cognitive functioning, to the extent that such do exist, are likely to take a long time to become manifested. Short-term studies of the kind described earlier show perhaps what effects *can* be produced under controlled conditions, but not what *actually* happens in psychological, social, and cultural reality.

In contrast to demonstrating the possibility of short-term, educationally induced cognitive consequences of technology, the possibility of unintended long-term, generalizable effects of "naturally occurring interaction" with technology is still to be proven. This possibility raises at least three major questions. First, what is the time scale along which such effects can be observed to take place on individuals' minds, let alone on a whole culture, the way Havelock (1963) has studied the societal consequences of literacy. Second, given that relevant observations of such effects are possible, to what extent do the effects actually take place? Is there any empirical evidence to lend support to the hypothesis that technology leads to *effects of* in any broad and lasting way? Third, what in technology, which of its elements or functions, would be expected to generate such effects?

EFFECTS THROUGH METAPHORS

Having compared and contrasted *effects with, effects of,* and *effects through,* we turn to one final example—the curious case of metaphorical models. It is a familiar point that we recruit metaphors from the concrete side of life and language for thinking about the abstract world (Lakoff and Johnson, 1980). Metaphors of mind are among the notable examples. As mentioned earlier, some technologies offer new metaphors to think about mind with—"the brain as a computer" (Bolter, 1984)—taking a step further along the path from the earlier concept of mind as a mechanical device, the clockwork mind. In the same spirit, statistical tools are said to lead to the development of psychological theories (Gigerenzer, 1991). Gigerenzer (1991) advanced an extended argument to the effect that statistical tools such as ANOVA offer new theoretical metaphors that radically change the kind of phenomena observed, recorded, and interpreted in psychology.

Such examples present a provocative case of the impact of technology on cognition: new and powerful metaphors to think with. It is worth asking how this case classifies into the framework of *effects with, effects of,* and *effects through.* It is interesting to note that analogical models appear to be a kind of hybrid case, presenting some features of *effects through* and some features of *effects of.* As to *effects through,* new metaphors of mind certainly are transformative. They do not simply extend and refine the way we think about mind, but rather, generate a substantially reorganized activity system of explanation. At the same time, such metaphors exhibit a prime characteristic of *effects of*: One does not need to be using the technology at the time to benefit. Indeed, one does not even have to be deeply expert with the technology. Discussing the mind-as-computer certainly requires some familiarity with how computers and programming work, but not the expertise of a professional programmer or systems engineer. One only needs to know enough to exercise the metaphor generatively. Accordingly, the example of metaphors of mind shows that the framework of *effects with, effects of,* and *effects through* is best applied flexibly, not just as a set of three bins into which everything must classify neatly, but in the more nuanced spirit of three perspectives for appraising how cognitive technologies impinge on the complex life of the mind.

So, back to the question we began with, "Does technology make people smarter?" or, more formally, "Do technologies expand our cognitive capabilities in any fundamental sense?" The answer offered here is assuredly yes. However, it is a nuanced yes rather than a broad and unqualified one. First, *cognitive technologies* are most at issue, technologies that directly accomplish cognitive functions. To be sure, other technologies—for instance, those concerned with health or communications—undoubtedly influence cognition, but this is not the interesting issue in the present context. Second, in many cases, cognitive capability must be interpreted in the person-plus sense of the person with tools as a system. Although one might object, "That's not really smarter," we do well to remember that human beings are normally, not exceptionally, tool inventors and users and the best measures of human accomplishment need to recognize that.

With these points in mind, at least three kinds of effects can be discerned—*effects with* technology, amplifications of cognitive capability as the technology is used; *effects of,* residual effects without the technology that is due to substantial experience with it; and *effects through,* effects largely with the technology that go beyond simply enhancement to a fundamental reorganization of the cognitive activity in question. The three are quite different in their dynamics: Initial *effects with* generally emerge relatively rapidly and prove substantial, but develop into true expertise only for some assiduous practitioners; *effects of* are relatively small compared to the magnitude of *effects with* and develop gradually over time; and *effects through* emerge gradually as individuals and societies explore the full affordances of the technology in question.

REFERENCES

Bolter, D. J. (1984). *Turing's man. Western culture in computer age.* Chapel Hill: University of North Carolina Press.

Bolter, D. J. (1991). *Writing space: The computer, hypertext, and the history of writing.* Hillsdale, NJ: Lawrence Erlbaum Associates.

Bruner, J. S. (1966). *Studies in cognitive growth.* New York: Wiley.

Chandler, P., & Sweller, J. (1991). Cognitive load theory and the formats of instruction. *Cognition and Instruction, 8,* 293–332.

Cline, H. F., & Mandinach, E. (1994). *Classroom dynamics: Implementing a technology-based learning environment.* Hillsdale, NJ: Lawrence Erlbaum Associates.

Cole, M., & Griffin, P. (1980). Cultural amplifiers reconsidered. In D. R. Olson (Ed.), *The social foundations of language and thought: Essays in honor of Jerome S. Bruner* (pp. 343–365). New York: Norton.

Cothey, V. (2002). A longitudinal study of World Wide Web users' information searching behavior. *Journal of the American Society for Information Science and Technology, 53,* 67–82.

Daiute, C. (1985). *Writing and computers.* Reading, MA: Addison-Wesley.

Ellul, J. (1964). *The technological society.* New York: Knopf.

Entwistle, N. (1996, April). *Conceptions of learning and understanding.* Paper presented at the American Educational Research Association annual meeting, New York.

Gigerenzer, G. (1991). From tools to theories: A heuristic of discovery in cognitive psychology. *Psychological Review, 98,* 254–267.

Green, C. S., & Bavelier, D. (2003). Action video game modifies visual selective attention. *Nature, 423,* 534–537.

Greenfield, P. (1972). Oral and written language: The consequences for cognitive development in Africa, the United States and England. *Language and Speech, 15,* 169–178.

Grotzer, T. A., & Perkins, D. N. (2000). Teaching intelligence: A performance conception. In R. J. Sternberg (Ed.), *Handbook of intelligence* (pp. 492–515). New York: Cambridge University Press.

Havelock, E. (1963). *Preface to Plato.* Cambridge, MA: Harvard University Press.

Hutchins, E. (1995). How a cockpit remembers its speed. *Cognitive Science, 19,* 265–288.

Lakoff, G., & Johnson, M. (1980). *Metaphors we live by.* Chicago: University of Chicago Press.

Luria, A. R. (1976). *Cognitive development: Its cultural and social foundations.* Cambridge, MA: Harvard University Press.

Olson, D. R. (1986). Intelligence and literacy: The relationships between intelligence and the technologies of representation and communication. In R. J. Sternberg & R. K. Wagner (Eds.), *Practical intelligence: Nature and origins of competence in the everyday world* (pp. 338–360). New York: Cambridge University Press.

Pea, R. (1993). Practices of distributed intelligence and designs for education. In G. Salomon (Ed.), *Distributed cognitions: Psychological and educational considerations* (pp. 111–138). New York: Cambridge University Press.

Perkins, D. N. (1985). The fingertip effect: How information-processing technology changes thinking. *Educational Researcher, 14*(7), 11–17.

Perkins, D. N. (1993). Person plus: A distributed view of thinking and learning. In G. Salomon (Ed.), *Distributed cognitions* (pp. 88–110). New York: Cambridge University Press.

Perkins, D. N., & Ritchhart, R. (2004). When is good thinking? In D. Y. Dai & R. J. Sternberg (Eds.), *Motivation, emotion, and cognition: Integrative perspectives on intellectual functioning and development* (pp. 351–384). Mahwah, NJ: Lawrence Erlbaum Associates.

Perkins, D. N., Tishman, S., Ritchhart, R., Donis, K., & Andrade, A. (2000). Intelligence in the wild: A dispositional view of intellectual traits. *Educational Psychology Review, 12*(3), 269–293.

Ritchhart, R., & Perkins, D. N. (in press). Learning to think: The challenges of teaching thinking. In K. Holyoak (Ed.), *Cambridge handbook of thinking and reasoning.* New York, NY: Cambridge University Press.

Ritchhart, R. (2002). *Intellectual character: What it is, why it matters, and how to get it.* San Francisco: Jossey-Bass.

Salomon, G. (1993a). *Distributed cognitions: Psychological and educational considerations.* New York: Cambridge University Press.

Salomon. G. (1993b). No distribution without individuals' cognition: A dynamic interactional view. In G. Salomon, *Distributed cognitions: Psychological and educational considerations.* New York: Cambridge University Press.

Salomon, G. (1994). *Interaction of media cognition and learning.* Mahwah, NJ: Lawrence Erlbaum Associates. (Original work published 1979)

Salomon, G., Perkins, D. N., & Globerson, T. (1991). Partners in cognition: Extending human intelligence with intelligent technologies. *Educational Researcher, 20,* 2–9.

Scribner, S., & Cole, M. (1981). *The psychology of literacy.* Cambridge, MA: Harvard University Press.

Trageton, A. (2001). Creative writing on computers grades 1–4. httpi//ans.hsh.no/home/utr/tekstskaping/artiklor/creativewritingoncomputersGrade2.htm. Accessed 22 October, 2003.

Wolfram, S. (2002). *A new kind of science.* Champaign, IL: Wolfram Media.

5

Cognitive Tools for the Mind: The Promises of Technology—Cognitive Amplifiers or Bionic Prosthetics?

Susanne P. Lajoie
McGill University

Technology has often been touted as an agent of educational change. However, the likelihood of technology fostering such change is greatly increased when cognitive theories guide design. Unfortunately, far too often, sweeping statements are made as to the possibilities for technology to lead to improvements in classrooms and learning in general. To my mind technology is a tool, a means to an end. Tools are designed for a purpose and their effectiveness can only be assessed within the context of that purpose. One theme in the educational literature that most exemplifies my own approach to the study of cognition using technology is that of *cognitive tools* (Jonassen & Reeves, 1996; Kommers, Jonassen, & Mayes, 1992; Lajoie, 2000; Lajoie & Derry, 1993; Pea, 1985; Perkins, 1985; Salomon, Perkins, & Globerson, 1991). The metaphor implies that there are technological tools that can assist learners to accomplish cognitive tasks. The question addressed in this chapter is whether such cognitive tools amplify the mind, serve as bionic prostheses, or do something completely different.

A COGNITIVE TOOLS APPROACH

I agree that technology can amplify what we know, but I support Pea's (1985) assumption that cognitive tools go beyond amplification and can help learners reorganize their knowledge in a manner that results in deeper understanding. The cognitive tools theme goes beyond the promotion of acquiring more knowledge through the use of technology. For example, just as access to hard copies of the *Encyclopedia Britannica* does not guarantee that the owner of the volumes understands or knows everything in those volumes, access to the Internet does not guarantee more and better knowledge. In fact, Salomon et al. (1991) suggest a more interesting approach to the use of technology whereby computers serve as intellectual partners that help learners accomplish tasks. The notion of a partnership implies that there is a sharing of information that leads to positive outcomes. Sharing, by definition, means that one person or thing, that is, a computer, does not do all the work.

Richard Snow once asked whether the term *cognitive tools* could be interchanged with cognitive or *bionic prosthetics.* By definition, bionic refers to the use of an electronic device to replace damaged limbs, organs, or functions (Encarta® World English Dictionary). Hence, a cognitive bionic prosthetic device implies replacing something that is broken or missing. Cognitive tools could support skills that were missing, but they are not designed based on a deficit model or intended to replace "parts" or "functions."

However, such prosthetics may become a reality considering that a symbionic mind movement (Cartwright & Finkelstein, 2002) is in place that pertains to the design and development of brain–computer interfaces. Some things that were once considered science fiction are now science, such as research on the development of emgors (electromyogram sensors that enable amputees to control artificial limbs in an almost natural manner), cerebella stimulators (brain pacemakers that prevent deep depression and epileptic seizures, and reduce intractable pain) and virtual reality. At first read of these advances you might think that "The Matrix" is reality and the reality that we live in is artificial. However, 20 years ago we never would have thought that you could tell your phone to call your husband, and it would automatically call him, nor would you imagine walking into your house and telling the lights to come on through voice activation. Perhaps thought activation will be the next technological advance, where implanted bionic chips can add memory when needed. Bionic prosthetics may become reality sooner than we think. However, for the purpose of this chapter cognitive tools pertain to what is available in the present.

Cognitive tools aid cognition through interactive technologies that expand the mind (Jonassen, 1996; Jonassen & Reeves, 1996; Kommers, Jonassen, & Mayes, 1992; Lajoie, 1993, 2000; Lajoie & Derry, 1993; Pea, 1985; Perkins, 1985; Salomon, et al., 1991). Cognitive tools help students during thinking, problem solving, or learning by providing them with opportunities to practice applying their knowledge in the context of complex, meaningful activities rather than in isolation of their ultimate use. For example, labeling types of bacteria may not be as meaningful as solving patient cases where bacteria is part of the disease.

Another type of cognitive tool is one that frees up a learner's resources by performing lower-level operations. By performing the smaller tasks, there are still processing resources left over for the learner to perform higher-order tasks. The premise is that lower-order skills may slow down the resources available for the overall problem-solving activity. For example, if learners spend all their time doing mathematical calculations, they might not have time to reflect on the meaning of the data that they calculate.

Probably the most powerful cognitive tool is one that allows learners to generate and test hypotheses in the context of complex problem solving. For instance, medical students who have not worked with human patients could practice their diagnostic reasoning with simulated patients, thereby having opportunities to safely engage in cognitive activities that would be out of their reach until later in the program of study. Tools can be designed to scaffold learning by providing diagnostic feedback based on computer assessment of learners' misconceptions or errors during problem solving.

Other cognitive tools are designed to support specific cognitive processes (e.g., memory, metacognition). For example, graphical or textual representations that summarize learners' problem-solving processes, be it evidence collection or data analyses, assist in making learners independent self-assessors who can reflect on what they have already done and on how their own problem-solving skills compare to cognitive components that have been identified as indicators of proficiency. Video scenarios can also be designed to serve as exemplars for learners.

Cognitive tools can be designed to represent what learners know or to allow learners to represent their knowledge in multiple ways. In either case, multiple forms of representation maximize opportunities for reaching individual differences in learning. Cognitive tools can also support cooperative learning within a problem-based learning environment.

Throughout the rest of this chapter I use one learning environment, BioWorld, as a context for elaborating on the cognitive tools themes. The theoretical framework and the rationale for following this approach are discussed, as are the tools and the type of data that can be collected to make inferences regarding the strength of the cognitive tools approach. The chapter concludes with a section on future research endeavors.

COGNITIVE TOOLS IN ACTION: BIOWORLD

For the past several years, my research has focused on designing cognitive tools that promote learning and assessment in science classrooms. These cognitive tools are packaged in BioWorld, a computer-based learning

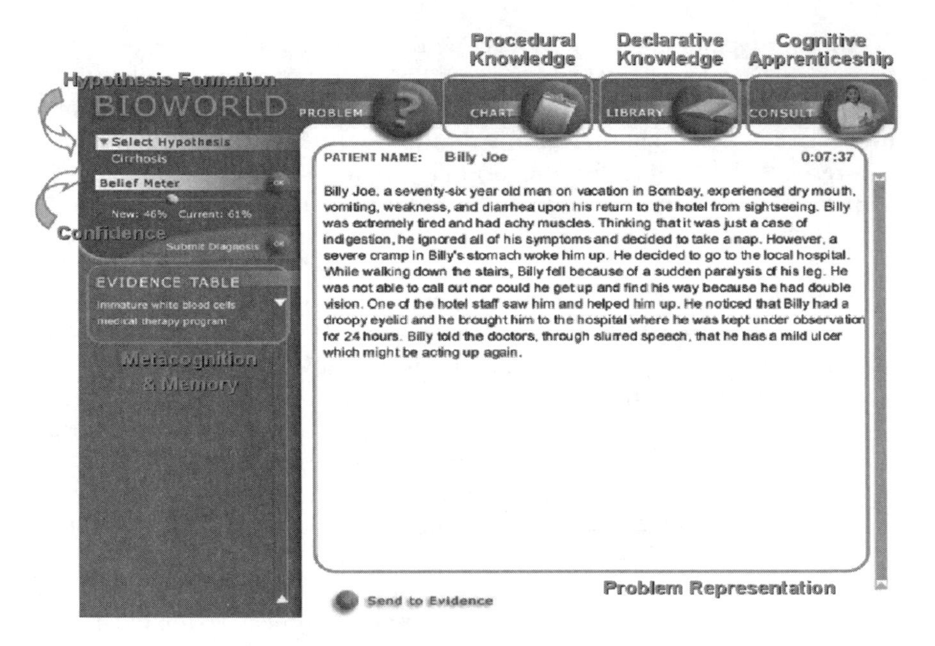

Fig 5.1. The Current BioWorld Interface: Case History Screen.

environment that increases students' scientific reasoning in the context of problem-based learning situations (Lajoie, Lavigne, Guerrera, & Munsie, 2001). BioWorld serves to instruct, model proficiency, and assess knowledge.

BioWorld is now implemented on the Web, allowing for platform independence. The same cognitive tools are present now as in previous versions (see Fig. 5.1). BioWorld complements the biology curriculum by providing a hospital simulation where students can apply what they have learned about body systems to problems in which they reason about diseases. This environment attempts to achieve a closer correspondence between classroom learning and real-world applications of science (Collins, 1997; National Academy of Sciences, 1994).

Each problem starts with a patient case history and students formulate a hypothesis about the disease. Once students select a hypothesis, they indicate how confident they are about it using the Belief Meter (% certainty). Students collect evidence from the patient case that supports their hypothesis, and this evidence remains visible on the Evidence Palette. There is an online library where students access declarative knowledge about the disease they are researching. Information in the library represents the symptoms, diagnostic tests, and transmission routes of a specific disease, as well as a glossary of medical terminology. In order to solve problems, students must conduct diagnostic tests to confirm or disconfirm their hypothesis. They do so by ordering tests on the patient chart, where the outcomes of their tests are recorded. This chart is a procedural knowledge tool because it provides a way for actions to be conducted in the context of problem solving. A simulated consultation tool is present and labeled as a cognitive apprenticeship tool because learners can obtain feedback during the data collection process.

In line with the previous discussion, BioWorld falls under the cognitive tools approach in several ways. First, it assists students in complex problem-solving activities pertaining to scientific reasoning by engaging them in hypothesis formation and testing rather than teaching them isolated science skills such as memorizing biology vocabulary or categorizing anatomical parts. They are provided with opportunities to enter and change their hypotheses in the context of solving the realistic patient cases that they would not have opportunities to do without these tools. Furthermore, the Belief Meter provides a trace of how student beliefs about their hypotheses change or do not change throughout the problem-solving activity. Students engage in sophisticated higher-order reasoning skills while the computer performs lower-order skills. In this example, students conduct diagnostic tests that help confirm or disconfirm their hypotheses. At the same time, BioWorld interprets these tests as falling into a normal or abnormal range, saving students processing time in interpreting such data.

There are several tools in BioWorld that scaffold the learner in the context of problem solving. One obvious tool is the online library that provides students with a mechanism to acquire new declarative or factual knowledge about diseases. Another cognitive tool is the Evidence Palette, where students post the evidence they see as relevant to the case. This tool serves two purposes: one, it is an external memory device that reminds them of what they have collected, and two, it serves as a metacognitive tool in that it helps students self-assess and monitor information that they saw as relevant to a particular problem-solving scenario. Finally, the Consult button provides assistance to learners whenever help is requested.

Tools that are not demonstrated in Fig. 5.1 are the Evidence Categorization and Argumentation tools. After students enter a final diagnosis, their evidence is presented to them in a list. They are asked to first categorize the evidence by type and then to formulate a final argument that consists of a rank order of the evidence that was most relevant to their final solution. They then compare their final argument to an expert's list that provides oral summation of how to solve the patient case. These tools provide opportunities to identify relevant from irrelevant information, which is a necessary component of expertise. Furthermore, the tools help them self-assess and compare their performance with those more proficient than themselves.

WHAT ARE THE ADVANTAGES OF A COGNITIVE TOOLS APPROACH?

Why this approach and not some other? The cognitive tools approach is not really one approach but a combination of various theory-based approaches to the design of computer-based instruction. Every cognitive tool has an underlying purpose, and its design may be based on different theories. Thus, the cognitive tools themes are not restricted to one paradigm but rather are selective in choosing the model that fits the purpose of instruction or training. Hence, one must understand the context behind the tools and their design.

BioWorld was designed to answer questions about specific learning processes in the context of scientific reasoning and hypothesis generation. A major strength is that through computer tracing of student actions one can test which situations are providing opportunities for learning. For example, where do students spend the most time, or what do they do when they are in a particular situation? More specifically, when visiting the library, are they examining references pertinent to their hypotheses (for example, shigellosis) or are they reading material that is of general interest (e.g., gonorrhea, a disease that does not share any commonalities with shigellosis). Tools that promote the acquisition of self-monitoring skills have the ultimate goal of promoting life-long learning skills, or learning to learn new things. Another advantage of the tools designed for this system is that they provide a mechanism for looking at how knowledge evolves in parallel with self-confidence about that knowledge through the Belief Meter and Hypothesis Generation menus.

There are a number of issues that could be strengthened in this approach. One is a thorough plan for assessing the transfer of knowledge from one situation to another. In other words, how generalizable is this approach? Any cognitive tools approach needs to address how to assess complex reasoning patterns using technology. Although complex reasoning patterns can be identified through computer traces of actions, there is still a need to get at the underlying reasons behind specific actions. New tools might be designed for this purpose. Although the current approach speaks to the sociology of learning, there is still a need for better articulation of assessment models for the individual learner as well as for the group or community of learners sharing the tools. Hence, a future direction is to broaden these theories in a manner that can help operationalize these dual facets of learning.

What Does the Cognitive Tools Approach Have in Common With Other Approaches?

The cognitive tools approach relates to a number of approaches in the research community. For instance, microworlds often provide opportunities for hypothesis generation and testing. Microworlds are usually computer simulations where students work with existing variables, insert their own data, and test their hypotheses. Microworlds provide endless opportunities for hypothesis testing and reflection on the answers (de Jong & van Joolingen, 1998; Shute & Glaser, 1990). Similarly, problem-based learning (PBL) approaches (Barrows, 1986; Hoffman & Ritchie, 1997), when joined with technology, could represent some of the cognitive tools themes. PBL approaches present real-life problems that provide meaningful, challenging, and rich learning experiences that are more demanding cognitively (Koschmann, 1994) because they are more open-ended, complex, and require sustained periods of investigation (Schauble, Glaser, Duschl, Schulze, & John, 1995).

Similarly, project-based learning (Barron et al., 1998) and case-based reasoning (Schank, 1998) approaches can require sustained periods of investigation and support cognitive tools themes when coupled with technology-based instruction. The terms *cognitive tools* and *mind tools* are sometimes used interchangeably (Jonassen, 1996). Multimedia approaches often embed cognitive tools for learning with the premise that multiple representations provide opportunities for expressing learning and understanding through multiple modalities (Kozma, 2003; Mayer & Moreno, 2002; Mayer, Heiser, & Lonn, 2001).

The social aspects of tool usage are supported in multiple approaches that theoretically fit under the umbrella of situated learning and social constructivism, for example: communities of learners (COL; Brown, 1994, 1997); cognitive apprenticeship; and problem-based learning. The COL model refers to engaging students in meaningful research activities where each student has a role in the community with a goal of learning. This same approach can be extended to communities of practice (Barab & Duffy, 2000; Wenger, 1999) such as medicine or law, where students in the professions work together to solve cases. The social sharing of information in this approach is essential to the final outcome of the task. However, not all COL approaches use technology as a cognitive tool, but, as mentioned, some cognitive tools are used in social settings or distributed between students where collaboration is needed.

The cognitive apprenticeship model (Brown, Collins, & Duguid, 1989; Collins, Brown, & Newman, 1989; Gott, 1989; Lave & Wenger, 1991) provides a template for connecting abstract and real-world knowledge by creating new forms of pedagogy based on a model of learners, the task, and the situation in which learning occurs. There are four aspects of the pedagogical model—namely, content, methods, sequence, and sociology. Once again, researchers who use this model do not always embed cognitive tools technology within their designs. However, when they do embed technology, it can take on many of the cognitive tools themes. For example, content knowledge can be modeled for learners by providing them with examples of strategies used by competent learners. Pedagogical methods such as coaching and scaffolding can be used by computer coaches as well as human coaches (Lajoie, Faremo, & Wiseman, 2001). Sequencing instruction can also be controlled via technology based on student models of performance that would provide the computer tools to adapt to individual differences. The sociology of a cognitive apprenticeship model refers to situating learners in the context of a complex task shared by a learning culture.

What Theories Underlie the Cognitive Tools Approach?

The last decade revealed some shifts in what can be considered guiding metaphors for learning theory (Anderson, Greeno, Reder, & Simon, 2000; Anderson, Reder, & Simon, 1998; Brown, 1994; Greeno, 1998; Mayer, 1997). There was considerable debate as to whether or not situated and constructivist learning theories were more complete than existing cognitive theories that fell under the umbrella of information processing. Mayer's view is that theories evolve rather than replace one another, and hence, there are elements from each theory that remain and some that continue to be refined in an attempt to advance our understanding of learning. From the Anderson, Reder & Simon (1998) perspective, we need to refine our definitions of learning in social contexts. They purport that the social learning situation must be analyzed by studying the mind of each individual in that situation and how each individual contributes to the interaction. Greeno (1998) also advocates that new methods of interactivity and design experiments will lead to better definitions of learning. Research on this front began in the form of design experiments (Cobb, Stephan, McClain, & Gravemeijer, 2001). The cognitive tools and data described in BioWorld provide concrete examples of how situating learning in meaningful contexts provides new opportunities for understanding the dynamic nature of learning processes.

Learning theories are increasingly more inclusive in that cognition, motivation, and the social context in which learning takes place are considered as interconnected (Cordova & Lepper, 1996; Lepper, 1988). Examining learning processes as they occur within *situations* or meaningful contexts becomes more interesting than studying learning products. The dynamic nature of learning and the assessment of learning in progress is increasingly becoming a focus of research (Bransford & Schwartz, 1999; Lajoie, 2003; Lajoie & Lesgold, 1992; Mislevy, Steinberg, Breyer, Almond, & Johnson, in press; Pellegrino, Chudowsky, & Glaser, 2001).

Theories pertaining to the nature of expertise are pertinent to my research because many of the cognitive tools that are designed for specific learning environments are based on these theories (Chi, Glaser, & Farr, 1988; Ericsson, 2002; Glaser, Lajoie, & Lesgold, 1987; Lajoie, 2003). The assumption is that identifying dimensions of expertise can lead to improvements in instruction that will ultimately result in helping learners become more competent. Many of the approaches previously described, for example, communities of learning and cognitive apprenticeship models, have theories of expertise underlying instruction. For instance, in order to have a community of learners, knowledge must be shared. Further, some learners are more knowledgeable than others on certain things and, hence, become experts who can help others. Self-monitoring and metacognition are key elements of expertise; however, we need to identify what experts monitor in a specific context before we can assist novices in monitoring their behavior. Theories of individual differences are at the heart of all multimedia designs because these theories state that there are differences in the way students learn and process information. Two people may be equally intelligent but process information in different ways and have different learning preferences (Cronbach & Snow, 1977; Eisner, 1993; Snow, 1989). Implicit in the cognitive tools approach is that theories guide the design of cognitive tools based on the specific learning context.

EVIDENCE THAT COGNITIVE TOOLS ARE EFFECTIVE

A number of studies have been carried out with Grade 9 biology students in promoting scientific reasoning (Lajoie, Lavigne, Guerrera, & Munsie, 2001) and with Grades 7 to 9 learning disabled and nonlearning disabled students in a biology classroom (Guerrera, 2002; Guerrera & Lajoie, 1998). Some of these studies focus on the individual learner, whereas others compare small group learning situations where students work face-to-face and at a distance through the support of computer video conferencing (Lajoie, Guerrera, & Faremo, 1998). BioWorld promotes scientific reasoning with a range of student ability, including populations with learning disabilities.

In addition to looking at pre- and posttest data of learning gains, we are particularly interested in which cognitive tools are most conducive to learning and which features provide opportunities for learning. For this reason, student computer trace files are kept of all problem-solving activity and then analyzed. This data provides us with evidence of how systematic students are in their reasoning. For instance, are students thorough in the data they collect to confirm or disconfirm a hypothesis or do they randomly pick hypotheses until they stumble on the correct answer? Do students use the library to acquire all of the factual knowledge they need to know about their patient case? Do students use the knowledge to test a hypothesis, that is, perform the diagnostic tests associated with a specific disease? In addition to trace data, we audiotape students when they work in small groups, transcribe and code their protocols, and look for patterns of scientific reasoning. After coding such data, student plans are evident, as are their actions associated with their hypotheses. Using multiple forms of data to assess student learning provides a more complete picture of what students understand.

FUTURE ENDEAVORS

First and foremost, I am interested in whether or not the same cognitive tools designed in BioWorld generalize to new disciplines, such as medicine. The notion that cognitive tools might assist the spectrum of learners, from the learning disabled and nondisabled high school biology students (Guerrera, 2002) to medical students and physicians, is worth testing, as it will be a testimony to the robustness of the tool. Furthermore, testing a spectrum of learners will provide a model of emerging competence, in that scientific reasoning can be examined in different contexts. The answer will provide a more complex understanding of how CBLEs that provide the right mix of cognitive tools can promote learning across a developmental span of learning in the sciences and perhaps help us understand how trajectories of learning can be documented (Lajoie, 2003). Once these differences in learning are identified, different levels of scaffolding can be developed that can deepen understanding for learners across a knowledge continuum. To this end, BioWorld is currently being modified for research with medical students, residents, and staff physicians. We are just beginning to test medical students and residents using BioWorld cases to see if experts process information differently than novices (Faremo, Lajoie, Fleiszer, Wiseman, & Snell, 2001), and the results are encouraging. Differences have been found in the way different levels of medical personnel use the system. A related question is whether assessment using cognitive tools can result in a better understanding of what students know and, hence, to improvement in instruction based on this understanding. More specifically, can a methodology for assessing learning trajectories using BioWorld be designed and tested? In an attempt to improve instruction, a third question explores how human tutors assist small groups of learners. By documenting their instructional methods, it may be possible to inform the design of cognitive tools that provide better feedback to learners. By examining these situations, a model of effective scaffolding can be designed and used for subsequent learning and teaching situations with and without the use of technology. We have begun this process but more cases need to be examined prior to the development of a model of distributed tutoring (Lajoie, Faremo, & Wiseman, 2001).

A final question addresses the issue of how cognitive tools might be designed to promote distributed learning online. Although Internet access and use of information on the Internet are commonplace, theory-guided instruction using the Internet is less prevalent. This final question addresses the development of new cognitive tools for helping learners share knowledge and construct knowledge within their own learning community. The initial design of these tools will be for medical students, but the intent is that the conceptualization of the use of such tools can be applied in multiple contexts.

Expert medical tutors have both the content knowledge and the instructional methods that can be used to support learning in this online community. Medical small group instruction lends itself to a cognitive apprenticeship approach (Williams, 1992) whereby more expert-peers can guide learners through complex problem-solving activities. A cognitive apprenticeship model requires the systematic identification of the cognitive components of expertise or proficiency within a particular problem domain prior to the development of appropriate instructional methods. Not all tutors or teachers are expert pedagogues (Shulman, 1986). Identifying the content and methods used by experts can facilitate the instruction of subject matter experts and practitioners who may not have such pedagogical knowledge. Modeling pedagogical knowledge becomes a research question in itself (Sternberg & Horvath, 1995). Hence, part of this new research endeavor is to identify a model of pedagogical knowledge in this new domain. Results from prior analyses of a small group medical tutoring situation (Lajoie, Faremo, & Wiseman, 2001) will be used to design an appropriate Web-based instructional context.

Advancement in this research in the medical domain will first require a replication of the methodologies used in earlier studies. Medical students of varying levels of expertise will be compared with medical experts. From this data a model of the medical domain knowledge will be constructed. The development and validation of models of expertise is a necessary step in any research exploring the cognitive dimensions of instructional situations. Because domain-specific expertise is usually acquired in specific kinds of learning situations, any model of learning or development of expertise will have to take into account the nature of the situations of instruction. Concurrent to the replication studies, medical communities of practice (small group instruction with medical tutors) will be observed in an effort to inform the redesign of BioWorld to more accurately reflect learning in the workplace, not just the classroom.

CONCLUSION

A definition of the cognitive tools approach was provided along with commentary regarding the difference between a cognitive tool and a bionic prosthetic device. Cognitive tools could support skills that are missing, but they are not designed based on a deficit model or intended to replace parts or functions. A cognitive tools approach goes beyond amplifying what individuals know and helps learners reorganize their thinking and acquire new understanding. Describing the cognitive tools designed for BioWorld provided a concrete example of this approach in terms of its strengths and weaknesses. The overlap between the cognitive tools approach and other instructional approaches that either use or do not use technology was discussed. The theories that underlie the design of cognitive tools were reviewed. Examples of the types of data that can be collected in learning environments using cognitive tools were also provided to help readers with their own plans for research in this area. Finally, a section describing future research was discussed.

There is a long history of tool use in education and, hence, such precedents should be considered as leading to the development of the computers-as-cognitive-tools theme (Lesgold, 2000). Lesgold concludes that cooperative efforts among teachers, educational researchers, and information scientists will lead to better and more innovative tools that can be used with successful outcomes in classrooms. These tools take time to build, and the revision process is part of the development time. Cognitive tools can lead to effective learning environments that can serve as a platform for studying learning in a dynamic fashion as well as assessing learners in specific learning situations.

ACKNOWLEDGMENTS

Send Correspondence to Susanne P. Lajoie, 3700 McTavish St., Faculty of Education, McGill University, Montreal Quebec Canada H3A 1Y2. Research reported in this chapter was made possible through funding provided by the following granting agencies: the Canadian Social Sciences and Humanities Research Council, Valorisation Recherche Quebec, McConnell Foundation, Network for Centres of Excellence, and Office of Learning Technologies. Many graduate students (former and current) have contributed to the work that is reported here. Special thanks to Gloria Berdugo, Janet Blatter, Andrew Chiarella, Lucy Cumyn, Sonia Faremo, Genevieve Gauthier, Claudia Guerrera, Nancy Lavigne, Susan Lu, Carlos Nakamura, Thomas Patrick, and Jeffrey Wiseman, M.D.

REFERENCES

Anderson, J., Greeno, J. G., Reder, L., & Simon, H. A. (2000). Perspectives on learning, thinking and activity. *Educational Researcher, 29*(4), 11–13.

Anderson, J., Reder, L., & Simon, H. (1998). Radical constructivism and cognitive psychology. In D. Ravitch (Ed.), *Brookings papers on education policy* (pp. 227–278). Washington, DC: Brookings Institute Press.

Barab, S. A., & Duffy, T. M. (2000). From practice fields to communities of practice. In D. H. Jonassen & S. M. Land (Eds.), *Theoretical foundations of learning environments* (pp. 25–55). Mahwah, NJ: Lawrence Erlbaum Associates.

Barron, B. J., Schwartz, D. L., Vye, N. J., Moore, A., Petrosino, A., Zech, L., Bransford, J. D., & The Cognition and Technology Group at Vanderbilt (1998). Doing with understanding: Lessons from research on problem- and project-based Learning. *Journal of Learning Sciences, 7*(3&4), 271–311.

Barrows, H. S. (1986). A taxonomy of problem-based learning methods. *Medical Education, 20,* 481–486.

Bransford, J. D., & Schwartz, D. L. (1999). Rethinking transfer: A simple proposal with multiple implications. *Review of Research in Education, 24,* 61–100.

Brown, A. L. (1994). The advancement of learning. *Educational Researcher, 23*(8), 4–12.

Brown, A. L. (1997). Transforming schools into communities of thinking and learning about serious matters. *American Psychologist, 52*(4), 399–413.

Brown, J. S., Collins, A., & Duguid, P. (1989). Situated cognition and the culture of learning. *Educational Researcher, 18,* 32–42.

Cartwright, G. F., & Finkelstein, A. (2002). Second decade symbionics and beyond. *Journal of Evolution and Technology, 8,* 1–16.

Chi, M. T. H., Glaser, R., & Farr, M. (1988). *The nature of expertise.* Hillsdale, NJ: Lawrence Erlbaum Associates.

Cobb, P., Stephan, M., McClain, K., Gravemeijer, K. (2001). Participating in classroom mathematical practices. *The Journal for the Learning Sciences, 10*(1&2), 113–163.

Collins, A. (1997). National science educational standards: Looking backward and forward. *The Elementary School Journal, 97*(4), 299–313.

Collins, A., Brown, J. S., & Newman, S. E. (1989). Cognitive apprenticeship: Teaching the craft of reading, writing, and mathematics. In L. B. Resnick (Ed.), *Knowing, learning, and instruction: Essays in honor of Robert Glaser* (pp. 453–494). Hillsdale, NJ: Lawrence Erlbaum Associates.

Cordova, D. I., & Lepper, M. R. (1996). Intrinsic motivation and the process of learning: Beneficial effects of contextualization, personalization, and choice. *Journal of Educational Psychology, 88*(4), 715–730.

Cronbach, L. J., & Snow, R. E. (1977). *Aptitudes and instructional methods: A handbook for research on interactions.* New York: Irvington.

de Jong, T., & van Joolingen, W. R. (1998). Scientific discovery learning with computer simulations of conceptual domains. *Review of Educational Research, 68*(2), 179–201.

Eisner, E. (1993). Forms of understanding and the future of educational research. *Educational Researcher, 22*(7), 5–11.

Ericsson, K. A. (2002). Attaining excellence through deliberate practice: Insights from the study of expert performance. In M. Ferrari (Ed.), *The pursuit of excellence in education* (pp. 21–55). Hillsdale, NJ: Lawrence Erlbaum Associates.

Faremo, S. (2004). *Examining medical problem solving in a computer-based learning environment.* Unpublished doctoral dissertation, McGill University, Montreal, Canada. Manuscript in preparation.

Faremo, S., Lajoie, S. P., Fleiszer, D., Wiseman, J., & Snell, L. (2001, May). *Examining medical problem solving in a computer-based learning environment.* Paper presented at the Annual Meeting of the American Educational Research Association, Seattle.

Glaser, R., Lajoie, S., & Lesgold, A. (1987). Toward a cognitive theory for the measurement of achievement. In R. R. Ronning, J. Glover, J. C. Conoley, & J. C. Witt (Eds.), *The influence of cognitive psychology on testing, vol. 3* (pp. 41–85). Hillsdale, NJ: Lawrence Erlbaum Associates.

Gott, S. P. (1989). Apprenticeship instruction for real world cognitive tasks. *Review of Research in Education, 15*, 97–169.

Greeno, J. (1998). The situativity of knowing, learning, and research. *American Psychologist, 53*(1), 5–26.

Guerrera, C. (2002). *Testing the effectiveness of problem-based learning with learning disabled high school biology students.* Unpublished doctoral dissertation, McGill University, Montreal, Canada.

Guerrera, C., & Lajoie, S. P. (1998, April). *Investigating student interactions within a problem-based learning environment in biology.* Paper presented at the annual meeting of the American Educational Research Association, San Diego, CA.

Hoffman, B., & Ritchie, D. (1997). Using multimedia to overcome the problems with problem-based learning. *Instructional Science, 25*, 97–115.

Jonassen, D. H. (1996). *Computers in the classroom: Mindtools for critical thinking.* Columbus, OH: Prentice-Hall.

Jonassen, D. H., & Reeves, T. C. (1996). Learning with technology: Using computers as cognitive tools. In D. H. Jonassen (Ed.), *Handbook of research for educational communications and technology* (pp. 693–719). New York: Simon & Schuster.

Kommers, P., Jonassen, D. H., & Mayes, T. (1992). (Eds). *Cognitive tools for learning.* Berlin: Springer.

Koschmann, T. D. (1994). Toward a theory of computer support for collaborative learning. *The Journal of the Learning Sciences, 3*, 219–225.

Kozma, R. B. (2003). Material and social affordances of multiple representations for science understanding. *Learning and Instruction, 13*(2), 205–226.

Lajoie, S. P. (1993). Computer environments as cognitive tools for enhancing learning. In S. P. Lajoie & S. J. Derry (Eds.), *Computers as cognitive tools* (pp. 261–288). Hillsdale, NJ: Lawrence Erlbaum Associates.

Lajoie, S. P. (Ed.). (2000). *Computers as cognitive tools (vol. 2): No more walls.* Mahwah, NJ: Lawrence Erlbaum Associates.

Lajoie, S. P. (2003). Transitions and Trajectories for Studies of Expertise. *Educational Researcher, 32*(8), 21–25.

Lajoie, S. P., & Derry, S. J. (Eds.). (1993). *Computers as cognitive tools.* Hillsdale, NJ: Lawrence Erlbaum Associates.

Lajoie, S. P., Faremo, S., & Wiseman, J. (2001). Tutoring strategies for effective instruction in internal medicine. *International Journal of Artificial Intelligence and Education, 12*(3), 293–309.

Lajoie, S. P., Guerrera, C., & Faremo, S. (1998, August). *Promoting Argumentation in Face-To-Face and in Distributed Computer-Based Learning Situations: Constructing Knowledge in the Context of Bioworld.* Presented at the annual meeting of the Cognitive Scince Society, Madison, WI.

Lajoie, S. P., Lavigne, N. C., Guerrera, C., & Munsie, S. (2001). Constructing knowledge in the context of BioWorld. *Instructional Science, 29*(2), 155–186.

Lajoie, S. P., & Lesgold, A. (1992). Dynamic assessment of proficiency for solving procedural knowledge tasks. *Educational Psychologist, 27*(3), 365–384.

Lave, J., & Wenger, E. (1991). Situated Learning: Legitimate peripheral participation. Cambridge, MA: Cambridge University Press.

Lepper, M. (1988). Motivational considerations in the study of instruction. *Cognition and Instruction, 5*(4), 289–309.

Lesgold, A. (2000). What are the tools for? Revolutionary change does not follow the usual norms. In S. P. Lajoie (Ed.), *Computers as cognitive tools (vol. 2): No more walls* (pp. 399–409). Mahwah, NJ: Lawrence Erlbaum Associates.

Mayer, R. E. (1997). Learners as information processors: Legacies and limitations of educational psychologies second metaphor. *Educational Psychologist, 31*(3/4), 151–161.

Mayer, R. E., Heiser, J., & Lonn, S. (2001). Cognitive constraints on multimedia learning: When presenting more material results in less understanding. *Journal of Educational Psychology, 93*(1), 187–198.

Mayer, R. E., & Moreno, R. (2002). Aids to computer-based multimedia learning. *Learning & Instruction, 12*(1), 107–119.

Mislevy, R. J., Steinberg, L. S., Breyer, F. J., Almond, R. G., & Johnson, L. (in press). Making sense of data from complex assessment. *Applied Measurement in Education.*

National Academy of Sciences. (1994). *National science education standards.* Washington, DC: National Academy Press.

Pea, R. D. (1985). Beyond amplification: Using the computer to reorganize mental functioning. *Educational Psychologist, 20,* 167–182.

Pellegrino, J., Chudowsky, N., & Glaser, R. (2001). *Knowing what students know.* Washington, DC: National Academy of Sciences.

Perkins, D. N. (1985). The fingertip effect: How information processing technology shapes thinking. *Educational Researcher, 14,* 11–17.

Salomon, G., Perkins, D. N., & Globerson, T. (1991). Partners in cognition: Extending human intelligence with intelligent technologies. *Educational Researcher, 20*(3), 2–9.

Schank, R. C. (1998). *Inside multi-media case based instruction.* Hillsdale, NJ: Lawrence Erlbaum Associates.

Schauble, L., Glaser, R., Duschl, R. A., Schulze, S., & John, J. (1995). Students' understanding of the objectives and procedures of experimentation in the science classroom. *The Journal of the Learning Sciences, 4*(2), 131–166.

Shulman, L. (1986). Paradigms and research programs in the study of teaching: A contemporary perspective. In M. C. Wittrock (Ed.), *Handbook of research on teaching* (pp. 3–36). New York: MacMillan.

Shute, V. J., & Glaser, R. (1990). A large-scale evaluation of an intelligent discovery world: Smithtown. *Interactive Learning Environments, 1,* 55–77.

Snow, R. E. (1989). Toward assessment of cognitive and conative structures in learning. *Educational Researcher, 18*(9), 8–14.

Sternberg, R. J., & Horvath, J. (1995). A prototype view of expert teaching. *Educational Researcher, 24*(6), 9–17.

Wenger, E. (1999). *Communities of practice: Learning, meaning, and identity.* New York: Cambridge University Press.

Williams, S. M. (1992). Putting case-based instruction into context: Examples from legal and medical education. *The Journal for the Learning Sciences, 2*(4), 367–427.

III

TECHNOLOGICAL PARTNERSHIPS AT WORK

6

Work in Progress: Reinventing Intelligence for an Invented World

Alex Kirlik
University of Illinois at Urbana-Champaign

> *We should then liken the environment not to a container or backcloth within or against which life goes on but rather to a piece of sculpture, or a monument, except in two respects: first, it is shaped not by one hand but by many; and second, the work is never complete. No environment is every fully created, it is always undergoing creation. It is, as it were, 'work in progress'.*
>
> —Tim Ingold (1992, pp. 39–56)

Many people today live and work in almost completely manufactured environments. We are creatures inhabiting, navigating, and selecting among niches in an ecology of technology, especially in the modern workplace. Given that concepts of human intelligence have historically drawn heavily on adaptation to the environment as an organizing theme (Neisser et al., 1996), the invented and continually reinvented nature of today's environment creates (or should create) a sizable challenge for reinventing the study of intelligence and the concept of adaptation itself. As Ingold (1992) noted, the exercise of human intelligence today nearly always occurs in a world continually "undergoing creation," and our connection to that world is more than ever mediated and augmented by tools and technologies. It is unclear whether traditional conceptions of intelligence and its measurement, grounded in unaided and passive adaptation to experimental "stimuli" or test items, will provide much insight into how people live and work in concert with tools and technologies, and within a continually reinvented world.

Given its theoretical grounding in adaptation, our understanding of intelligent functioning can surely be no greater than our understanding of the environments we inhabit, and the manner in which we are connected to them. Evidence is not hard to find suggesting that progress in intelligence research is being impeded as much by our limited understanding of the functionally relevant aspects of the environment, as it is by our limited understanding of the purely internal aspects of intellectual functioning. Systems theorists such as Gardner (1993) and Sternberg (1997), for example, have put forward multidimensional theories of intelligence, where the various dimensions can, in many cases, be understood to be selectively associated with particular contexts or environmental demands. Sternberg's theory even acknowledges the ability to select and shape one's own environment, in addition to passively adapting to the given environment, as central to successful intellectual functioning (also see Kirlik, 1998a, 1998b; Kirsh, 1995, 1996).

Of course, there also is mounting evidence concerning the intensive content, context, and cultural specificity of intelligence and performance, not only in psychology (e.g., Anderson, 1990; Chater & Oaksford, 1999; Sternberg & Wagner, 1994; Zhang & Norman, 1997), but also in anthropology (Schiffer, 2001) and cultural theory (e.g., Cole, 1996). Decomposing the context-dependent and independent aspects of intellectual functioning remains a major challenge and is almost surely to depend on more precise theories of environmental contexts. Recently, there is even evidence that some scientists studying perhaps the most fundamentally internal, genetic contributions to intelligence are coming to the view that a more thorough understanding of the environment is a crucial prerequisite to advancing theory:

> What is needed is a more careful analysis of environments. We have no taxonomy of environments at present and an understanding of the complexities of environment/genetic interactions depends on being equally precise about both factors. In actual studies, it is even more difficult to define and specify the functional environment. What will be needed is a multi-dimensional approach to an understanding of "environments." (Mandler, 2001, pp. 155–156)

As one who has spent his career studying technological work and workplaces, and creating approaches for both understanding and supporting cognition and performance through design, my goal in this chapter is to present what we have learned about how one might go about developing knowledge of the environmental dimensions of intelligent behavior. One reason that design-oriented studies of technological work may have something useful to say about environmental aspects of intelligence is that the design challenge entails reasoning about, or searching through, a space of technologically feasible environments in order to enhance intellectual functioning. The ability to reason effectively over various environmental designs is crucial because the quality of a resulting design exerts strong influences on learning and performance (e.g., Hollnagel, 2003; Kirlik & Bisantz, 1999). As such, psychological theories useful for guiding design must be concerned as much, if not more, with what I like to call *environmental differences* as with individual differences (Kirlik, 1995).

Such an emphasis marks our work off from, and thus complements, much traditional research on intelligent behavior, as the latter has largely focused on understanding intelligence as a context-free property, or properties, of a person and how individuals may differ with respect to these properties. Gaining insights into individual differences is one quite valuable role of intelligence research, and indeed, this work has provided a technology of intelligence testing useful for workplace selection (e.g., Schmidt & Hunter, 1998). However, research focused solely on finding the best, most general predictors of intelligent functioning, such as *g* or general intelligence, abstracts away all of what we know to be immensely important differences between various environments and places them under a single name, "the environment" (some exceptions do exist: see, e.g., Ackerman & Cianciolo, 2002; Sternberg & Wagner, 1994).

In this sense, one can see that the psychological construct called the environment in most psychometric research has long been little more than a useful and convenient fiction, akin to the economist's convenient fiction: the "rational man" (or woman). Both fictions, the first glossing environmental differences, the second glossing individual differences, are now straining under the weight of their many assumptions (e.g., see Sunstein, 2000, for the rise of "behavioral economics"). Seen in this light, it is hardly a coincidence to find that none of the most highly regarded textbooks in human factors, a field with deep roots in technology interaction and environmental design, even index the term "intelligence" (Kantowitz & Sorkin, 1983; Sanders & McCormick, 1993; Wickens, Gordon, & Liu, 1998). Given that intelligence research is now increasingly coming to embrace the problem of environmental differences, perhaps the time is ripe for research on technological work and design organized around these differences to aid in a reinvention of intelligence better suited to an invented world.

FORM FOLLOWS FUNCTION: WORKING INTELLIGENTLY WITH TOOLS

It is hardly a surprise that the majority of scientific studies of intelligence have made little, if any, reference to people working with tools and technology. Despite the many conceptualizations of intelligence viable today, nearly all share a deep, often unstated, assumption that what is being discussed touches on one, if not the "best" or "highest," of the qualities defining who we, *Homo sapiens*, are. Since at least the time of *The Republic*, in which Plato relegates *techne*, or the skills of technological artisans, to the lowest rung of his sociopolitical hierarchy, our best or highest qualities have been conceived, in Western thought at least, as those involving detached, intellectual activities (cf. Arendt, 1998; Hickman, 2003).

Yet, Kirlik (1995) and a number of *distributed cognition* theorists (e.g., Clark, 1997, 2003; Hutchins, 1995; Norman, 1988; Olson & Olson, 1991) have emphasized that cognitive science may have made a mistake in so intimately equating *detached* and *intellectual*, thus targeting a search for the functional seat of cognition solely within the brain or mind. As discussed previously, the constraining effects of restricting search in this way are also now coming to be felt quite concretely by those conducting intelligence research as well. Two examples highlighting the need to embrace the functional participation of the environment in intellectual functioning will hopefully illustrate the need to broaden this search.

Work in Progress Lesson 1: Physics

In his wonderfully researched and written biography of the late Nobel prize-winning physicist Richard Feynman, James Gleick relates an episode in which MIT historian Charles Wiener was conducting interviews with Feynman at a time when Feynman had considered working with Wiener on a biography. Gleick writes that Feynman, after winning the Nobel prize, began dating his scientific notes, "something he had never done before" (Gleick, 1992, p. 409). In one discussion with Feynman, "Weiner remarked casually that his new parton notes represented 'a record of the day-to-day work,' and Feynman reacted sharply" (p. 409). What was it about Weiner's comment that drew a sharp reaction from this great scientist? Did he not like his highly theoretical research described merely as "day-to-day work"?

No, and the answer to this question reflects, to me at least, something of Feynman's ability to have deep insights, not only into physics, but into other systems as well. Feynman's reaction to Wiener describing his notes as "a record" was to say: "I actually did the work on the paper." (Gleick, 1992, p. 409). To which an apparently uncomprehending Wiener responded, "Well, the work was done in your head, but the record of it is still here" (p. 409). One cannot fail to sense frustration in Feynman's retort: "No, it's not a *record*, not really. It's *working*. You have to work on paper, and this is the paper. Okay?" (p. 409, italics in the original).

My take on this interchange is that Feynman had a deep understanding of how his work was comprised of a dynamic, functional transaction (Dewey, 1896) between his huge accumulation of internal cognitive tools as well as his external cognitive tools of pencil and paper, enabling him to perform functions such as writing, reflecting on and amending equations, diagrams, and so on (cf. Donald, 1991; Vygotsky, 1929/1981). Most tellingly, note his translation from Weiner's description of the world in terms of physical form ("No, it's not a *record*, not really") into a description in terms of function ("It's working").

Why did Weiner have such a difficult time understanding Feynman? External objects such as Feynman's notes do, of course, exist as things, typically described by nouns. Yet, in our functional transactions with these objects, the manner in which they contribute to intelligent behavior requires that these things be understood in functional terms, that is, in terms of their participation in the operation of the closed-loop, human–environment system (cf. Monk, 1998, on "cyclic interaction"). Weiner, like so many engineering students through the ages, demonstrated apparent difficulty in making the transition from understanding and describing the external world primarily in terms of form (nouns) to viewing the world primarily in terms of function (verbs).

Work in Progress Lesson 2: Architecture

This next example, from my own work, illustrates a number of points. First, it provides a concrete illustration of how design, as a search across feasible environments, operates in a reverse logical direction to much of psychology, thereby providing a way of thinking about human–environment relations that complements what is typically learned through psychological research. Whereas the latter typically specifies an environment and then searches for an explanation of cognition and behavior, the former has a specified or desired behavior in mind, and searches across *environmental differences* to find a design that most effectively supports that behavior. In this particular case, the behavior of interest was safe and efficient occupant egress from the new high-rise office building that, at the time of writing, is now under construction as a replacement for the World Trade Center Seven (WTC-7) building destroyed on September 11, 2001. In early 2002, I participated in initial design meetings for this new building at the New York offices of the architectural design firm of Skidmore, Owings and Merrill to provide guidance on human factors and cognitive engineering issues.

One of the many technological issues we considered at these meetings concerned how to foster safe and efficient egress. In the United States, during emergencies elevators are normally not available to occupants—they are used and supervised by firefighters, although this policy is now being reconsidered.

We, therefore, focused on hallway, and most important, stairwell design. Notably, this was an exercise of creativity in a straightjacket, as no less than a

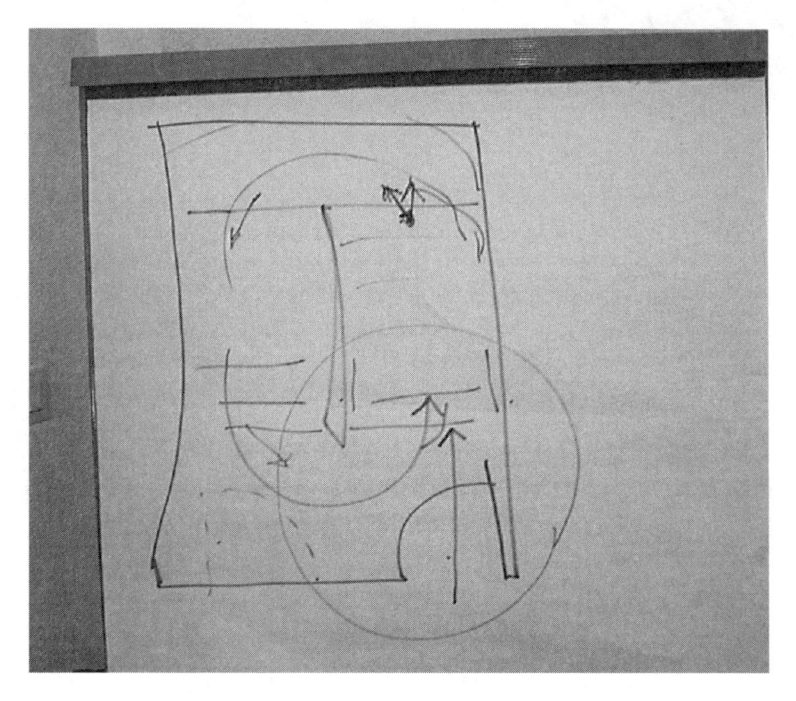

Fig 6.1. A solution to the doorway merge design problem to foster emergency egress.

dozen separate codes and standards govern various aspects of egress-related design for office buildings in the city of New York. Our own activities in technology design, thus, had to function in harmony with these codes and standards. Such are most fruitfully viewed as tools or technologies in their own right, because they embody previously determined prescriptions and adaptations (Bowker & Star, 2000).

Figures 6.1 and 6.2 depict work in progress on stairwell design, as photographed during work breaks by the author. Figure 6.1 represents a solution to the problem of where to place stairwell access doors so that flow through the doors merges in an orderly fashion with the flow of people already using the stairwell (no code specifies this). As in the previous example, work is literally done on paper: I am quite confident that our working group would more naturally call the design solution shown in Figure 6.1 "our work," as opposed to a "record of our work," and our dynamic transactions with the markers and tablets as literally "working." It is interesting to note that there are two levels of functionality illustrated by this diagram. First is the manner in which our working consisted of a functional transaction with the external representational medium (also see Suchman, 2000, on the role of artifacts in engineering design). Second, as we were designing to support effective human functioning, note how arrows were used in an attempt to animate the

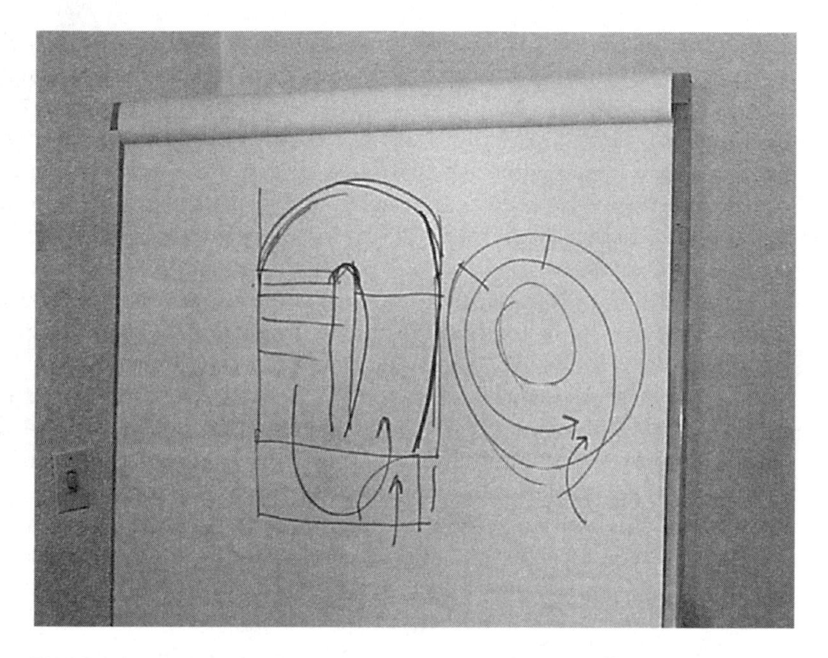

Fig 6.2. Attempts to solve the wasted kinetic energy problem in emergency egress.

functionality of the design in terms of a dynamic system in operation. Seeing the primary role of these diagrams as functional rather than formal comes across even more directly in Fig. 6.2, in which both arrows and a rounding of the stairwell corners have been drawn to indicate how a more circular flow of people down a stairwell would be more energy efficient, and thus more rapid. Standard square landings typically result in much wasted transfer of the body's kinetic energy at each landing because of the need to change direction 180 degrees in largely a stop-and-turn motion. However, a rounded design such as this was inconsistent with code, as tread widths are constrained to be uniform across the entire width of the tread. As such, we focused instead on increasing stair width beyond the traditional (code required) 44 inches, finding no scientific basis for this minimum width. In fact, we found scientific evidence suggesting that this width insufficiently supports effective egress in emergency situations, especially in cases where firefighters are climbing upward while occupants are moving downward (Pauls, 1985).

FUNCTIONING IN CONCERT WITH A FUNCTIONAL WORLD

Although the previous two examples of working with tools are largely anecdotal, I hope they have been at least somewhat persuasive in highlighting two themes that I believe will be of value in reinventing intelligence for a largely invented world. The first is the need to better understand the environment in functional terms, where the functionality described is tightly related to cognition and activity (see Kirsh, 2001, for a nice example in the context of office-based task management; and Woods, 2003, for a more general discussion of the need for functional analysis to understand cognitive systems in context). The second is the need to understand intelligence in terms of functional transactions with a functional world (e.g., van Geert, 2003). Neither of these themes is particularly new to psychology, as they were previously advanced in various forms by the early functionalists (e.g., William James, James Dewey) and ecologists (e.g., James Gibson, Egon Brunswik).

Yet, I nominate the resurrection of these themes to occupy a central role in psychological science and theory of intelligence, as I believe them to be essential for understanding intellectual functioning in a designed world awash with technology. The undeniable ubiquity of a world of people *using* tools and technologies is, I suggest, now doing what no crucial experiment or theory created by these ecologically oriented, functionalist pioneers could ever do. That is, to convince us that finely grained, functional analyses of particular environments are crucial prerequisites to achieving an adequate understanding of what it means, and what it takes, to function intelligently in concert with those environments. As noted earlier, and as demonstrated by the previous two examples, this analysis must also take into account any opportunities people have to participate in shaping their own environments using tools as well. Without detailed functional analysis, we have found, even the most basic task of defining what it means to function intelligently in particular contexts cannot be meaningfully approached.

In hopes of illustrating these points more concretely, I present an example of one such analysis and modeling effort using, in this case, a finely grained, functional description of an environment in terms of Gibson's (1979/1986) theory of affordances—a model of the environment in terms of opportunities for action. This study shed light onto the fluency of behavior in a highly complex, dynamic task; plausible explanations of the differences between high and low performers; and insights into why the knowledge underlying skill or expertise may, in some cases, appear to take on a tacit (Polanyi, 1966), or otherwise unverbalizable, form.

The *Scout World*: Modeling the Work Context With Dynamic Affordance Distributions

Consider Fig. 6.3, which depicts a participant in the experiment performing a dynamic, interactive simulation of a supervisory control task, described here as the *Scout World.* This laboratory research was motivated by the

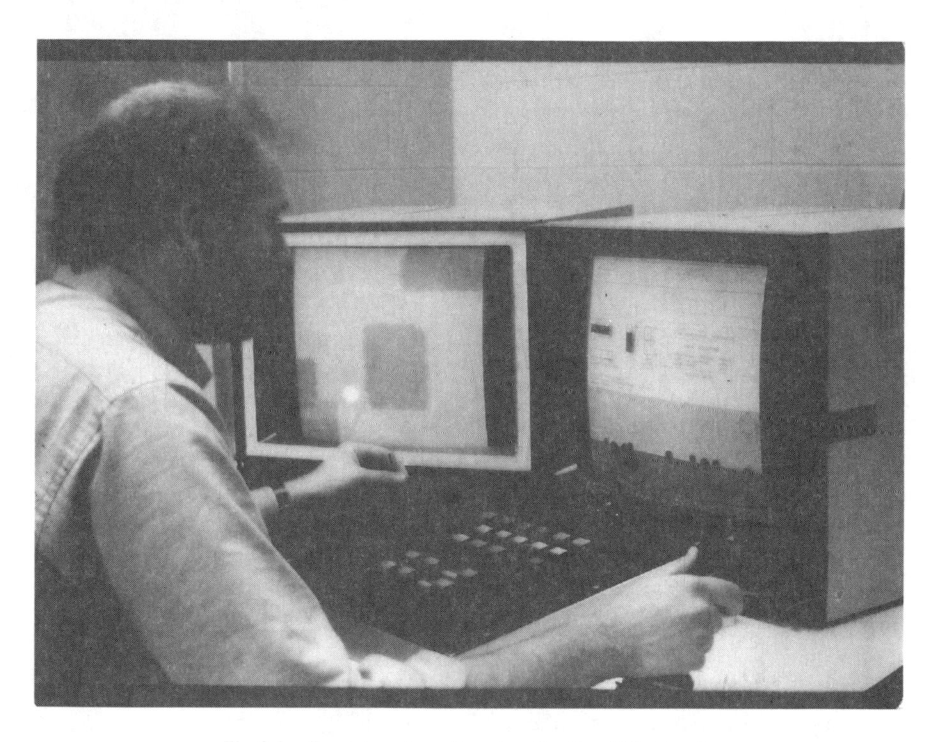

Fig 6.3. Participant performing the *Scout World* task.

practical question of whether a one- or two-person crew would be required to operate a future helicopter, and required the participant to control not only his or her own craft, called the Scout, but also four additional craft over which the participant exercised supervisory control (Sheridan, 1984) by entering action plans at a keyboard (e.g., fly to a specified waypoint, conduct patrol, load cargo, return to a home base, etc.). The left display in Fig. 6.3 depicts a top-down situation display of the partially forested, 100-square mile world to which activity was confined. The display on the right shows an out-the-window scene (lower half) and a set of resource and plan information for all vehicles under control (upper half). The participant's task was to control the activities of both the Scout and the four other craft to score points in each 30-minute session by processing valued objects that appeared on the display once sighted by Scout radar. See Kirlik, Miller, & Jagacinski (1993) for details.

Our goal was to create a computer simulation capable of performing this challenging task, and one that would allow us to reproduce, and thus possibly explain, differences between the performance of both one- and two-person crews, and novice and expert crews. At the time, the predominant cognitive modeling architectures, such as Soar (Newell, 1992), ACT-R (Anderson, 1990), and the like did not have mature perception and action resources allowing them to be coupled with external environments, nor had they been demonstrated to be capable of performing dynamic, uncertain, and interactive tasks (a limitation Newell agreed was a legitimate weakness of these approaches: see Newell, 1992). In addition, modeling techniques drawn from the decision sciences would have provided an untenably enumerative account of participants' decision processes and were rejected because of bounded rationality considerations (Simon, 1956).

Instead, we observed that our participants seemed to be relying heavily on the external world (the interface) as "its own best model" (Brooks, 1991). This was suggested not only by intimate perceptual engagement with the displays, but also by self-reports (by participants) of a challenging, yet engaging and often enjoyable sense of "flow" (Csikszentmihalyi, 1993) during each 30-minute session (not unlike any other "addictive" videogame or sport). We, thus, began to entertain the idea that if we were going to model the function of our human performers, we would have to model their world in functional terms as well if we were to demonstrate how the two functioned collectively and in concert. This turned us to the work of Gibson (1979/1986), whose theory of affordances provided an account of how people might be attuned to perceiving the world functionally; in this case, in terms of actions that could be performed in particular situations in the *Scout World.*

Following through on this idea entailed creating descriptions of the environment using the experimental participant's capacities for action as a frame of reference to achieve a functional description of the *Scout World* environment. A now classic example of this technique was presented by Warren (1984), who measured the riser heights of various stairs in relation to the leg lengths of various stair climbers and found, in this ratio, a functional invariance in people's ability to perceptually detect whether a set of stairs would be climbable (for them) or not. Warren interpreted this finding to mean that people could literally perceive the "climbability" of the stairs; that is, that people can perceive the world, not only in terms of form, but in functionally relevant terms as well.

Similarly, we created detailed, quantitative models of the *Scout World* environment in terms of the degree to which various environmental regions and objects afforded searching (discovering valued objects by radar), processing those objects (loading cargo, engaging enemy craft), and returning home to unload cargo and reprovision. Because participants' actions influenced the course of events experienced, they partially shaped the affordances of their own worlds.

Figure 6.4 contains a set of four maps of the same *Scout World* layout, including a representation purely in terms of visual form, as shown to participants (a), and functional representations in terms of affordances for

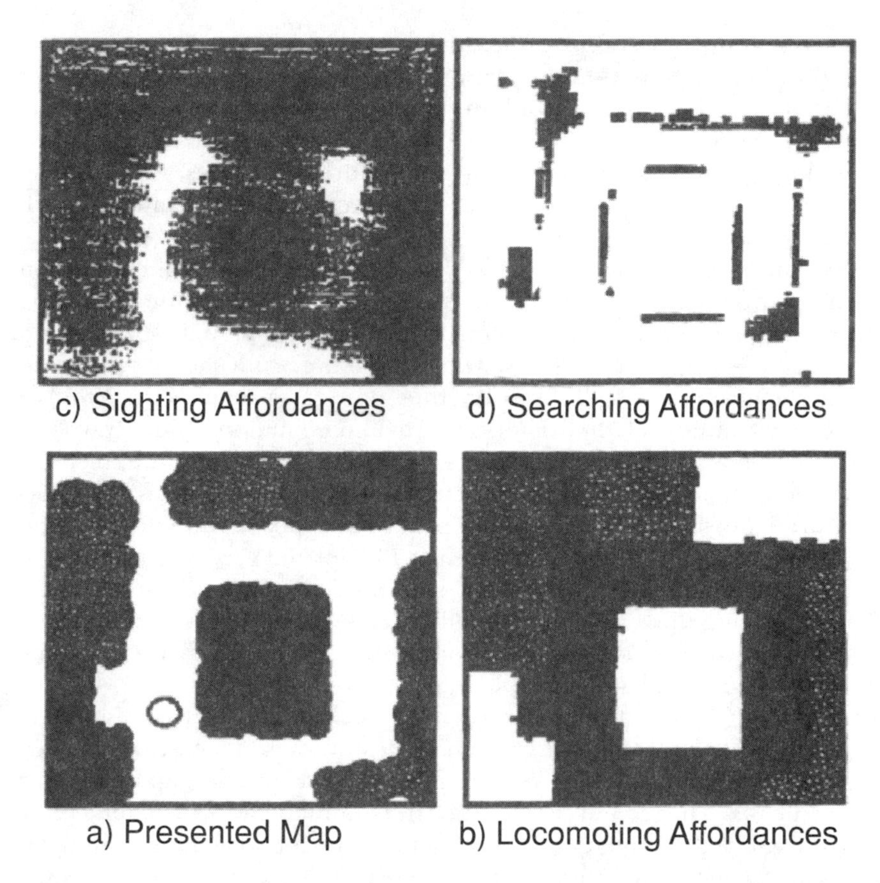

Fig 6.4. Four maps of the same Scout World layout: (a) The presented world map, (b) Map of affordances for locomotion, (c) Map of affordances for sighting objects, (d) Final searching affordance map.

actions of various types (b, c, d). For the Scout, for example, locomotion (flying) was most readily afforded in open, unforested areas (the white areas in Fig. 6.4a), and less readily afforded as forest density grew. As such, Fig. 6.4b shows higher locomotion affordances as dark and lower affordances as lighter. (Here we are simply using grayscale coding to represent these affordance values to the reader; in the actual model, the "dark" regions had high quantitative affordance values, and the "light" regions had relatively low quantitative affordance values.) Because the Scout radar for sighting objects (another action) had a 1.5 mile radius, and valued objects were more densely scattered in forests, the interaction between the Scout's capacity for sighting and the forest structure was more graded and complex, as shown in Fig. 6.4c (darker areas again indicating higher sighting affordance values).

Considering that the overall affordance for searching for objects was comprised of both locomotion and sighting affordances (searching was most readily afforded where one could most efficiently locomote and sight objects), the final searching affordance map in Fig. 6.4d was created by superimposing Figs. 6.4b and 6.4c (adding their affordance values, and rescaling for clarity). Figure 6.4d, thus, depicts ridges and peaks that maximally afforded the action of searching.

As explained in Kirlik et al. (1993), this affordance-based differentiation of the environment provided an extremely efficient method for mimicking the search paths created by participants. We treated the highest peaks and ridges on this map as successive waypoints that the Scout should attempt to visit at some point during the mission, thus, possessing an attractive *force*. Detailed Scout motion was then determined by a combination of these waypoint forces and the entire, finely graded, search affordance structure, or field. On the one hand, as one might expect, placing a heavy weight on the attractive forces provided by the waypoint peaks (as opposed to the entire field of affordances) resulted in Scout motion that looked very goal-oriented in its ignorance of the immediately local search affordance field. On the other hand, reversing these weights resulted in relatively meandering, highly opportunistic Scout motion that was strongly shaped by the local details of the finely grained search affordance field.

In an everyday situation such as cleaning one's house, the first case would correspond to rigidly following a plan to clean rooms in a particular order, ignoring items that could be opportunistically straightened up or cleaned along the way. The second case would correspond to having a general plan, but being strongly influenced by local opportunities for cleaning or straightening up as one moved through one's house. In the actual, computational *Scout World* model, this biasing parameter was set in a way that resulted in Scout search paths that best mimicked the degree of goal-directedness versus opportunism in the search paths observed.

For object-directed rather than region-directed actions, such as loading cargo or visiting home base, the *Scout World's* affordances were centered on those objects rather than distributed continually in space. As shown in Fig. 6.5, we created a set of dynamic affordance distributions for these discrete, object-directed actions for both the Scout and the four craft under supervisory control (F1–F4 in Fig. 6.5a). Each of the 15 distributions shown in Fig. 6.5a indicates the degree to which actions directed toward each of the environmental objects that can be seen in Fig. 6.5b were afforded at a given point in an action-based (rather than time-based) planning horizon. Space precludes a detailed explanation of how these distributions were determined (see Kirlik et al., 1993, for more detail). To take one example, consider craft F1, over which the participant had supervisory control by entering action plans via a keyboard. F1 appears in the northwest region of the world

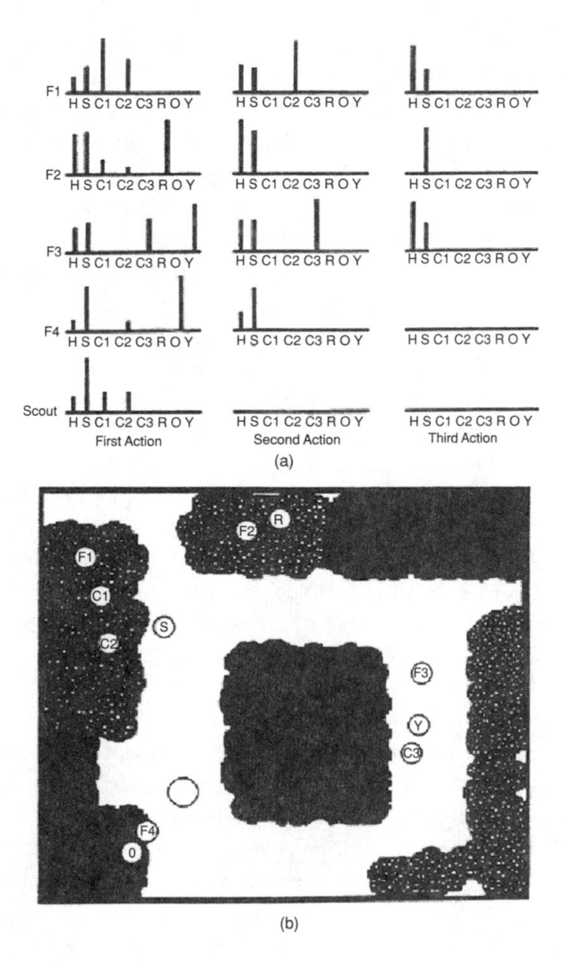

Fig 6.5. Two representations of the same world state: (a) functional representation in terms of dynamic affordance distributions; (b) representation in terms of visual form.

as shown in Fig. 6.5b, nearby is a piece of cargo labeled C1. The "First Action" affordance distribution for F1 indicates that this is the action most highly afforded for this craft, and a look down the column for all of the other craft, including the Scout, indicates that the affordance for loading this cargo is no higher for any craft other than F1. Thus, the model would, in this case, "decide" to assign the action of loading this piece of cargo to F1.

Given that F1 had been committed in this fashion, the model was then able to determine what the affordances for F1 would be at the time it had completed loading this cargo. This affordance distribution for F1 is shown in the "Second Action" column of distributions. Notice there is no longer any affordance for loading C1 (as this action will have been completed), and now the action of loading the cargo labeled C2 is most highly afforded. In this case, a plan to load this cargo allowed the model to generate a "Third Action" affordance distribution for F1, in this case indicating that the action of visiting home base H would be most highly afforded at that time, because of the opportunity to then score points by unloading two pieces of cargo.

What is crucial to emphasize, however, is that Fig. 6.5 provides a mere snapshot of what was actually a dynamic system. Just moments after the situation represented by this snapshot, an event could have occurred that would have resulted in a radical change in the affordance distributions shown (such as the detection of an enemy craft by radar). So, although I have spoken as if the model had committed to plans, these plans actually functioned solely as a resource for prediction, anticipation, and scheduling, rather than as prescriptions for action (cf. Suchman, 1987). The "perceptual" mechanisms in the model, tuned to measure the value of the environmental affordances shown in Figs. 6.4 and 6.5, could be updated 10 times per second, and the actual process of selecting actions was always determined by the affordances in the "First Action" distribution for all craft. Thus, even though the model would plan when enough environmental and participant-provided constraint on the behavior of the controlled system allowed it to do so, it abandoned many plans as well. A central reason for including a planning horizon in the model was to avoid conflicts among the four craft and the Scout: For example, "knowing" that another craft had a plan to act on some environmental object removed that object from any other craft's agenda, and "knowing" that no other craft's plans did not include acting on some other object increased the affordance for acting on that object for the remaining craft.

The components of the model intended to represent functions performed by internal cognition consisted of the previously mentioned perceptual mechanisms for affordance detection, and also a simple mechanism for combining the affordance measures with priority values keyed to the task payoff structure (e.g., points awarded per type of object processed). Notably, as described in Kirlik et al. (1993), these priority values turned out to be largely unnecessary because an experimental manipulation varying the task payoff structure (emphasizing either loading cargo or engaging enemy craft) by a ratio of 16:1 had *no* measurable effect on the behavior of participants. This finding lent credence to the view that participants' behavior was intimately tailored to the dynamic affordance structure of the *Scout World*, a set of opportunities for action that performers' actions themselves played a role in determining. Because of the fact that behavior involved a continual shaping of the environment, any causal arrow between the two would have to point in both directions (Dewey, 1896, Jagacinski & Flach, 2003). The general disregard of payoff information in favor of exploiting affordances is also consistent with the (or at least my) everyday observation that scattering water bottles around one's home is much more likely to prompt an increase of one's water consumption than any urging by a physician to do so.

Additionally, we manipulated the planning horizon of the model and found that the variance that resulted was not characteristic of expert–novice differences in human performance. This task apparently demanded less thinking-ahead than it did keeping-in-touch. In support of this view, what *did* turn out to be the most important factor in determining the model's performance, and a plausible explanation for expert–novice differences in this task, was the time required for each perceptual update of the world's affordance structure. As this time grew (from 0.5 s to 2 s), the model (and participants, our validation suggested) got further and further behind in their ability to opportunistically exploit the dynamic set of action opportunities provided by the environment, in a cascading, positive-feedback fashion. This result highlights that many, if not most, dynamic environments, or at least those we have studied, favor fast but fallible, rather than accurate but slow, methods for profitably conducting one's transactions with the world.

A final observation concerning our affordance-based modeling concerns the oft-stated finding that experts or skilled performers are notoriously unable to verbalize rules or strategies that presumably underlie their behavior. When shown a concrete situation or problem, in contrast, these same experts are typically able to report a solution with little effort. This phenomenon is often interpreted using constructs such as "tacit knowledge" (Polanyi, 1966) or "automaticity" (e.g., Shiffrin & Dumais, 1981). If one does assume, for the sake of discussion, that much procedural knowledge can usefully be described in terms of "if *p* then *q*" conditionals or rules, then our *Scout World* modeling provides a different explanation of why experts may often be unable to verbalize knowledge. Rather than placing such "if *p* then *q*" rules in the "head" of our model, we instead created perceptual mechanisms that functioned to "see" the world functionally, as affordances, which we interpret as playing the roles of the *p* terms in the "if *p* then *q*" construction. The *q*, on the other hand, is the internal response to assessing the world in functional terms, and as such, the "if *p* then *q*" construct is distributed across the boundary of the human–environment system. At least this was the case in our computational model.

As such, even if the capability existed to allow our model to introspect and report on its "knowledge," like human experts, it could not have verbalized any "if *p* then *q*" rules either, because it only contained the "then *q*" parts of these rules. If we instead "showed" the model any particular, concrete *Scout World* situation, it would have been able to readily select an intelligent course of action. Perhaps human experts and skilled performers have difficulty reporting such rules for the same reason: At high levels of skill or expertise, these conditionals, considered as knowledge, become distributed across the person–context system, and are, thus, not fully internal entities (cf. Greeno, 1987, on situated knowledge). Although speculative, perhaps the distribution of this knowledge across the human–environment system is also why our participants reported such an intimate sense of engagement or flow while performing at high levels of skill.

EJStars: Using Perceptual Augmentation to Support Affordance Detection

After our experiences modeling human performance in *Scout World,* it was natural to question whether we might be able to enhance the functional coupling between a performer and a functional environment by making affordances easier to perceptually detect.

Figure 6.6, for example, depicts a laboratory simulation of a display used by controllers of unpiloted aerial vehicles, or UAVs (Kirlik, Walker, Fisk, & Nagel, 1996). In this study, we showed that the introduction of novel perceptual cues (shading in the upper-right portion of the display), better

Fig 6.6. The perceptually augmented *EJStars* UAV display.

specifying the functional constraints on task performance, both accelerated skill acquisition and also fostered an increased level of protection against increased workload. The design of this augmentation resulted from a study of the functional constraints a performer had to respect to reach a high level of achievement and, subsequently, creating integrated, graphical forms to communicate this functional information directly. In a baseline version of this interface, a performer would have been required to mentally integrate disparate cues to infer this functional information. Our augmentation aided the performer by offloading this inference task from the performer to the display, by integrating information within the display itself. As such, we moved this task out of the head and into the world.

Using Tools and Action to Shape One's Own Work Environment

In the *EJStars* example we demonstrated that, as designers, we could lessen the cognitive burdens of a task by offloading some cognitive demands to the world. In some cases, however, we find that workers use tools and actions to do this job for themselves. For example, Kirlik (1998a, 1998b) presented a field study of short-order cooking, showing how more skilled cooks used strategies for placing and moving meats to create novel, and functionally reliable, information sources unavailable to cooks of lesser skill. We observed a variety of different cooks using three different strategies to ensure that each piece of meat (hamburgers) placed on the grill was cooked to the specified degree of doneness (rare, medium, or well).

The simplest ("brute force") strategy observed involved the cook randomly placing the meats on the grill and using no consistent policy for moving them. As a result, this cook's external environment contained relatively little functionally relevant information. The second ("position control") strategy we observed was one where the cook placed meats to be cooked to specified levels at specified locations on the grill. As such, this strategy created functionally relevant perceptual information useful for knowing how well each piece of meat should be cooked, thus eliminating the demand for the cook to keep this information in internal memory. Under the most sophisticated ("position + velocity control") strategy observed, the cook used both an initial placement strategy as well as a dynamic strategy for moving the meats over time. Specifically, the cook placed meats to be cooked well done at the rear and rightmost section of the grill. Meats to be cooked medium were placed toward the center of the grill (back to front) and not as far to the right as the meats to be cooked well done. Meats to be cooked rare were placed at the front and center of the grill. Interspersed with his other duties (cooking fries, garnishing plates, etc.), this cook then intermittently slid each piece of meat at a relatively fixed rate toward the left border of the grill, flipping them about halfway in their journey across the grill surface. Using this strategy, everything that the cook needed to know about the task was perceptually available from the grill itself, and, thus, the meats signaled their own completion when they arrived at the grill's left boundary.

In order to abstract insights from this particular field study that could potentially be applied in other contexts (such as improving the design of frustratingly impenetrable information technology), we decided to model this behavioral situation formally "to abstract away many of the surface attributes of work context and then define the deep structure of a setting" (Kirsh, 2001, p. 305). To do so, we initially noted that the function of the more sophisticated strategies could perhaps best be understood, and articulated, as creating constraints or correlations to exist between the value of environmental variables that could be directly observed and thus considered *proximal*, and otherwise unobservable, covert, or *distal* variables. As such, we were drawn to consider Brunswik's theory of probabilistic functionalism, which represents the environment in terms of exactly these functional proximal–distal relations (Brunswik, 1956; Hammond & Stewart, 2001). These ideas are articulated in Brunswik's lens model, shown in Fig. 6.7.

Brunswik advanced the lens model as a way of portraying perceptual adaptation as a "coming to terms" with environment, functionally described as probabilistic relations between proximal cues and a distal stimulus. As illustrated in Hammond & Stewart (2001), this model has been quite influential in the study of judgment, where the cues may be the results of medical observations and tests, and the judgment (labeled "Perception" in Fig. 6.7) is the physician's diagnosis about the covert, distal state of a patient (e.g., whether a tumor is malignant or benign). In our judgment research, we have extended this model to dynamic situations (Bisantz et al., 2000), and also to tasks in which cognitive strategies are better described by rules or heuristics rather than by statistical (linear regression-based) strategies (Rothrock

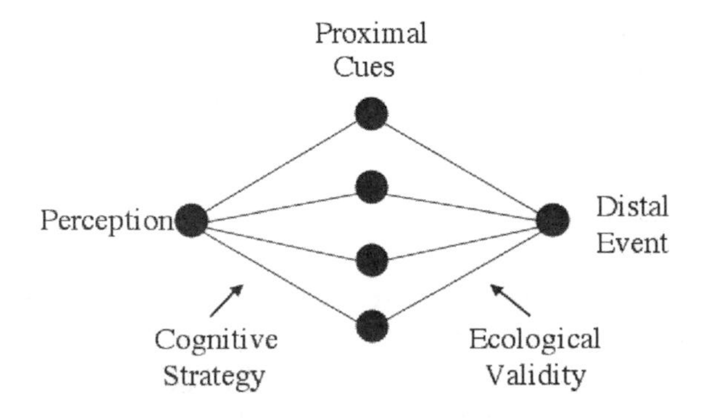

Fig 6.7. Brunswik's lens model.

& Kirlik, 2003). Note that the lens model represents a distributed cognitive system, where half the model represents the external proximal–distal resources supporting adaptation, and the other half represents the internal strategies or knowledge by which adaptation is achieved by using these resources.

Considering the cooking case, one deficiency of the lens model should become immediately apparent: In its traditional form, it lacks resources for representing the proximal–distal structure of the environment for action, that is, the relation between proximal means and distal ends or goals. The conceptual precursor to the lens model, originally developed by Tolman & Brunswik (1935), actually did place equal emphasis on proximal–distal functional relations in both the cue–judgment and means–ends realms. As such, we sought to extend the formalization of at least the *environmental* components of the lens model to include both the proximal–distal structure of the world of action, as well as the world of perception and judgment. The structure of the resulting model is shown in Fig. 6.8.

This extended model represents the functional structure of the environment, or what Brunswik termed its "causal texture," in terms of four different classes of variables, as well as any lawful or statistical relationships among them, representing any structure in the manner in which they may covary. The first [PP,PA] variables are proximal with respect to both perception and action: Given an agent's perceptual and action capacities, their values can be both directly measured and manipulated (in Gibson's terms, they are directly perceptible affordances). Variables [PP,DA] can be directly perceived by the agent but cannot be directly manipulated. Variables [DP,PA], on the other hand, can be directly manipulated but cannot be directly perceived. Finally, variables [DP,DA] can be neither directly perceived nor manipulated. Distal inference or manipulation occurs through causal links with proximal variables, in exactly the manner discussed by Tolman and Brunswik (1935).

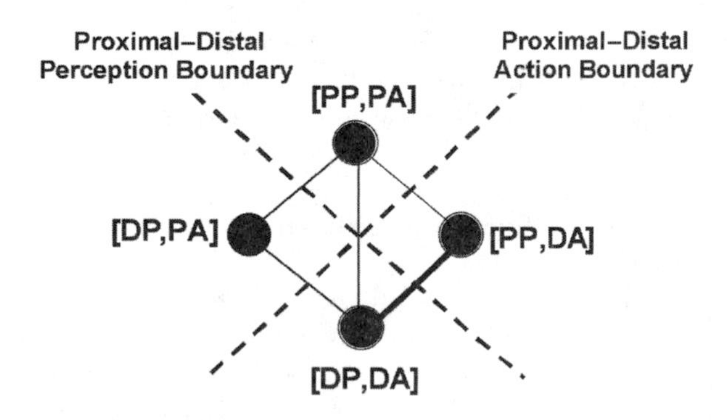

Fig 6.8. A functional model of the environment for perception and action.

Note the thick link between the variables [PP,DA] and the variables [DP,DA]. These two variable types, and the single relation between them, are the only elements of environmental structure that appear in the traditional lens model depicted in Fig. 6.7. All of the additional model components and relations represented in Fig. 6.8 have been added to be able to represent both the perceptual and action structure of the environment in a unified system. See Kirlik (1998a) for a more complete presentation.

To formally analyze the cooking case, we used this model to describe whether each functionally relevant environmental variable (e.g., the doneness of the underside of a piece of meat) is either proximal (directly perceivable; directly manipulable) or distal (must be inferred; must be manipulated by manipulating intermediary variables) under each of the three cooking strategies observed. Entropy-based measurement (multidimensional information theory: see McGill, 1954, for the theory, see Kirlik, 1998a, for the application to the cooking study) revealed that the most sophisticated cooking strategy rendered the dynamically controlled grill surface itself its "own best model" (Brooks, 1991), thereby allowing cooks to offload memory demands to the external world.

Quantitative modeling revealed that the most sophisticated (position + velocity) strategy resulted in, by far, the greatest amount of variability or entropy in the proximal, perceptual variables in the cook's ecology. This variability, however, was tightly coupled with the values of variables that were covert, or distal to other cooks, and, thus, this strategy had the function of reducing the uncertainty associated with this cook's distal environment nearly to zero. More generally, we found that the demands this workplace task placed on internal cognition were *underdetermined* without a precise, functional analysis of the proximal and distal status of both perceptual information and affordances, along with a functional analysis of how workers used tools to adaptively shape their own cognitive ecologies. This research also demonstrated that useful insights can be achieved by marrying the otherwise largely independent Brunswikian and Gibsonian schools of ecological psychology: See Gibson (1957/2001), Cooksey (2001), and Kirlik (2001) for a dialogue on the relationship between Brunswik's and Gibson's approaches, and Kirlik (1995) for a framework viewing them as complementary.

Modeling the Origins of Taxi Errors at Chicago O'Hare

Figure 6.9 depicts an out-the-window view of the airport taxi surface in a high-fidelity NASA Ames Research Center simulation of a fogbound Chicago O'Hare airport. The pilot is currently in a position where only one of these lane lines constitutes the correct route of travel. Taxi navigation errors, and especially errors known as runway incursions, are a serious threat to aviation

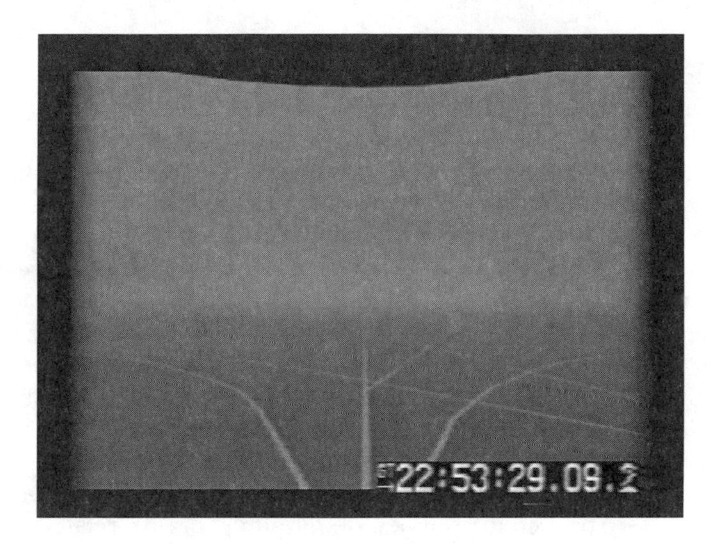

Fig 6.9. Simulated view of the Chicago O'Hare taxi surface in fog (Courtesy of NASA Ames Research Center).

safety. As such, NASA has pursued both psychological research and technology development in an effort to reduce these errors and mitigate their consequences. In my recent collaborative research with Mike Byrne, we completed a computational modeling effort aimed at understanding why experienced airline flight crews may have committed particular navigation errors in the NASA simulation of taxiing under these foggy conditions (for more detail on the NASA simulation and experiments, see Hooey and Foyle, 2001; for more detail on the computational modeling, see Byrne & Kirlik, 2003, in press).

Notably, the resulting model was comprised of a dynamic, interactive simulation, not only of pilot cognition, but also of the external, dynamic visual scene, the dynamic taxiway surface, and a model of aircraft (B-767) dynamics. In the task analyses with subject matter experts (working airline captains), we discovered five strategies pilots could have used to make turn-related decisions in the NASA simulation:

1. Accurately remember the set of clearances (directions) provided by air traffic control (ATC) and use signage to follow these directions.
2. Derive the route from a paper map, signage, and what one can remember from the clearance.
3. Turn in the direction of the destination gate.
4. Turn in the direction that reduces the maximum of the X or Y (cockpit-oriented) distance between the aircraft and destination gate.
5. Guess.

Our model was built around the idea that the selection of a decision strategy for each particular possible turn was determined by a cost–benefit tradeoff in which the most accurate decision strategy was chosen given the time horizon available (Payne & Bettman, 2001). Through extensive functional analysis of the environment, data provided by NASA, and Monte Carlo modeling, we concluded that pilots would have most likely used the "smart heuristics" (Raab & Gigerenzer, in press), represented by strategies 3 and 4 when decision horizons were between 2s and 8s, which, because of visibility conditions, were not atypical of many horizons in the simulation. In addition, we found that these simple heuristics were deceptively accurate, the latter resulting in a decision accuracy above 90%. Furthermore, an examination of the NASA error data revealed that a total of 12 taxi navigation errors were committed. Verbal transcripts indicated that 8 of these errors involved decision making, whereas the other 4 errors involved flight crews losing track of their location on the airport surface (these "situation awareness" errors were beyond the purview of our model of turn-related decision making). Consistent with our modeling, every one of the 8 decision errors in the NASA data set involved either an incorrect or premature turn toward the destination gate. Finally, we found that at *every* simulated intersection in which the instructed clearance violated *both* heuristics, at least one decision error was made. In these cases, the otherwise functionally adaptive strategies used by pilots to navigate under low visibility conditions steered them astray, because of an atypical combination of clearance and taxiway geometry.

CONCLUSION: BEYOND THE INTERFACE GIVEN

An increasing number of intelligence researchers are now coming to embrace the idea that a more encompassing, context-sensitive, and practically relevant understanding of intelligent behavior requires paying as much attention to environmental differences as is currently paid to understanding individual differences. In a largely invented world, our research suggests that a promising path toward developing a more sophisticated understanding of the often intimate and highly varied ways in which the environment participates in intelligence will require a sustained series of investigations into the detailed, functional structure of the human ecology.

Our research also suggests that reinventing intelligence may require a reorientation of intelligence away from its historical focus on passive adaptation. Instead, intelligent behavior may be better understood to be in the spirit of selecting and effectively trading leads with a dance partner; that is, the ability to conduct individually and socially productive functional transactions with the world. I hope the studies presented here will result in at least a few useful insights into how empirical and theoretical research organized

Fig 6.10. Prototype HUD taxi aid (Courtesy of NASA Ames Research Center).

around functional analyses of ecological niches, and the functional transactions people conduct with them, may aid in reaching these goals.

In closing, I would like to acknowledge, and indeed highlight, the other side to this story. An increasingly technological world also makes new sorts of demands on intelligence. To illustrate, consider the design of the "head up display" (HUD) NASA is currently evaluating as a possible way of reducing the frequency of taxi navigation errors and runway incursions, as shown in Fig. 6.10. A display such as this, like the UAV display shown in Fig. 6.6, adds proximal cues or perceptual augmentation (such as the STOP sign at center) to improve pilots' abilities to detect the functional constraints on their behavior. However, it is axiomatic that technology designers can never anticipate every environmental contingency (Vicente, 1999). Thus, they can never be certain that any such display (and its associated instrumentation) will provide sufficient information to specify *every* functional constraint on behavior *all* of the time, just as the pilots we modeled could never be certain that their experiential rules-of-thumb would *never* be defeated by an atypical combination of an ATC clearance and local taxiway geometry.

Fig 6.11. A delivery truck on the Chicago O'Hare Airport surface as photographed by the author. Can *every* functional constraint on behavior be anticipated in technology design?

In this light, one can view a display such as this HUD in exactly the same *functional* terms as one can view the taxi decision heuristics that pilots had developed through their own encounters with the taxi environment, or the building codes we encountered in our WTC-7 design work. All are, in a literal sense, technologies, the products of human mind and hand. Although generally supportive, they must always be viewed as potentially fallible (consider the eight taxi errors committed, the fact that better supporting emergency egress required us to design to a standard above and beyond code and tradition, that the policy against building occupants using elevators for egress is now undergoing reconsideration, and Fig. 6.11, with caption).

My own answer to the question posed in the caption of Fig. 6.11 is no. Technologies such as interfaces, automation, codes, and procedures are crystallized anticipations or adaptations, typically providing support in mediating or augmenting our interactions with a distal world. However, because of our inherent inability to predict the entire range of functionally relevant constraints in an environment, technology is always capable of creating a cognitive demand for the user to go beyond the interface, rule, or tool given. This going beyond requires one to understand the anticipations embodied in tool or technology design, and whether they are relevant to the current purpose. The need for a technology user to be able to learn, notice, and act on this distinction has also been noted by Dourish (2001), who described this cycle in terms of "engagement," "separation," and "re-engagement" (p. 141). Research on cognitive flexibility and metacognition, such as that presented by Dekker (2003), Jamieson & Miller (2000), Reder & Schunn (1999), and Frensch & Sternberg (1989), may provide useful insights on this issue.

As the cognitive ecology becomes ever more artificial, people will be increasingly required to adopt a reflective, and even critical, attitude toward technology, and monitor and evaluate their mediated coupling to the environment. When should I let my (or my culture's) tools, rules, and other technologies do my work for me, and when should I question, amend, or even override them? When should I be working transparently through technology, and when should I focus instead on the crystallizations embodied in tools and technologies as the proximal objects of my reason and action? Adaptively answering these questions for oneself is likely be a key to productively functioning in concert with an invented world. This conclusion is supported by a wealth of technology interaction research highlighting the difficulty of fostering an effective coupling between people and automation (for an overview, see Parasuraman & Riley, 1997, on "Humans and Automation: Use, Misuse, Disuse, Abuse"; and for a functional modeling example see Kirlik, 1993). After many years observing just how difficult it can be to design and seamlessly deploy technologies into work contexts in a truly supportive way, I have come to the conclusion that a major factor is a vast underestimation of the complexity, intimacy, and nuance of the functional coupling between people and their worlds. This problem will not go away until we have much more powerful and sophisticated ways of thinking about this coupling.

Behaving intelligently in an invented world requires the ability to selectively and effectively dance with, and dance around, technology in an individually and socially productive way. We are now, indeed, awash in tools. Intelligence is the handle that fits them all.

ACKNOWLEDGMENTS

Support for writing and the aviation safety research was provided by NASA Ames Grant NAG 2-1609 to the University of Illinois, David Foyle, technical monitor. I used the WTC-7 design problem as a teaching example in a seminar on applied ecological psychology while a guest of the Center for the Ecological Study of Perception and Action at the University of Connecticut. I thank the members of that seminar, especially Claudia Carello, Claire

Michaels, Theo Rhodes, Mike Richardson, and Jeff Wagman, for their participation, and Mike Byrne for his collaboration on the taxi modeling. I also thank Anna Cianciolo, Kim Vicente, and the editors for valuable comments on a previous version.

REFERENCES

Ackerman, P. L., & Cianciolo, A. T. (2002). Ability and task constraint determinants of complex task performance. *Journal of Experimental Psychology: Applied, 8*(3), 194–208.

Anderson, J. R. (1990). *The adaptive character of thought.* Hillsdale, NJ: Lawrence Erlbaum Associates.

Arendt, H. (1998). *The human condition,* 2nd ed. Chicago: University of Chicago Press.

Bisantz, A., Kirlik, A., Gay, P., Phipps, D., Walker, N., & Fisk, A. D. (2000). Modeling and analysis of a dynamic judgment task using a lens model approach. *IEEE Transactions on Systems, Man, and Cybernetics, 30*(6), 605–616.

Bowker, G. C., & Star, S. L. (2000). Invisible mediators of action: Classification and the ubiquity of standards. *Mind, Culture, and Activity,* 7(1&2), 147–163.

Brooks, R. (1991). Intelligence without representation. *Artificial Intelligence, 47,* 139–159.

Brunswik, E. (1952). The conceptual framework of psychology. In *International Encyclopedia of Unified Science, Vol. 1* (No. 10, pp. 1–102). Chicago: University of Chicago Press.

Brunswik, E. (1956). *Perception and the representative design of psychological experiments.* Berkeley, CA: University of California Press.

Byrne, M., & Kirlik, A. (2003). *Using computational cognitive modeling to diagnose possible sources of aviation error* (Tech. Rep. No. AHFD-03-14/NASA-03-04). Retrieved Jan 1, 2004 from University of Illinois, Aviation Human Factors Division Web Site: http://www.aviation.uiuc.edu/UnitsHFD/report_fulltext.html

Byrne, M., & Kirlik, A. (in press). Using computational cognitive modeling to diagnose possible sources of aviation error. *International Journal of Aviation Psychology.*

Chater, N., & Oaksford, M. (1999). Ten years of the rational analysis of cognition. *Trends in Cognitive Sciences, 3*(2), 57–64.

Clark, A. (1997). *Being there: Putting brain, body and world together again.* Cambridge, MA: MIT Press.

Clark, A. (2003). *Natural-born cyborgs.* New York: Cambridge University Press.

Cole, M. (1996). *Cultural psychology: A once and future discipline.* Cambridge, MA: Harvard University Press.

Cooksey, R. W. (2001). On Gibson's review of Brunswik and Kirlik's review of Gibson. In K. R. Hammond & T. R. Stewart (Eds.), *The essential Brunswik* (pp. 242–244). New York: Oxford University Press.

Csikszentmihalyi, M. (1993). *Flow: The psychology of optimal experience.* New York: HarperCollins.

Dewey, J. (1896). The reflex arc concept in psychology. *Psychological Review, 3,* 357–370.

Donald, M. (1991). *Origins of the modern mind: Three stages in the evolution of culture and cognition.* Cambridge, MA: Harvard University Press.

Dekker, S. (2003). Failure to adapt or adaptations that fail: Contrasting models on procedures and safety. *Applied Ergonomics, 34,* 233–238.

Dourish, P. (2001). *Where the action is: The foundations of embodied interaction.* Cambridge, MA: MIT Press.

Frensch, P. A., & Sternberg, R. J. (1989). Expertise and intelligent thinking: When is it worse to know better? In R. J. Sternberg (Ed.), *Advances in the psychology of human intelligence, Vol. 5* (pp. 157–188). Hillsdale, NJ: Lawrence Erlbaum Associates.

Gardner, H. (1993). *Multiple intelligences: The theory in practice.* New York: Basic Books.

Gibson, J. J. (2001). Survival in a world of probable objects: Review of Egon Brunswik (1956) perception and the representative design of psychological experiments. In K. R. Hammond & T. R. Stewart (Eds.), *The essential Brunswik* (pp. 244–246). New York: Oxford University Press. (Reprinted from *Contemporary Psychology, 2*(2), 33–45, 1957)

Gibson, J. J. (1986). *The ecological approach to visual perception.* Hillsdale, NJ: Lawrence Erlbaum Associates. (Original work published in 1979)

Gleick, J. (1992). *Genius: The life and science of Richard Feynman.* New York: Pantheon.

Greeno, J. G. (1987). Situations, mental models, and generative knowledge. In D. Klahr & K. Kotovsky (Eds.), *Complex information processing* (pp. 285–316). Hillsdale, NJ: Lawrence Erlbaum Associates.

Hammond, K. R., & Stewart, T. R. (Eds.). (2001). *The essential Brunswik.* New York: Oxford University Press.

Hickman, L. (2003). Doing and making in a democracy: Dewey's experience of technology. In R. C. Scharff & V. Dusek (Eds.), *Philosophy of technology: The technological condition* (pp. 369–377). Oxford, UK: Blackwell.

Hollnagel, E. (2003). *Handbook of cognitive task design.* Mahwah: NJ: Lawrence Erlbaum Associates.

Hooey, B. L., & Foyle, D. C. (2001). A post-hoc analysis of navigation errors during surface operations. Identification of contributing factors and mitigating strategies. *Proceeedings of the 11th Symposium on Aviation Psychology,* pp. 226:1–226:6.

Hutchins, E. (1995). *Cognition in the wild.* Cambridge, MA: MIT Press.

Ingold, T. (1992). Culture and the perception of the environment. In E. Croll & D. Parkin (Eds.), *Bush base: Forest farm, culture and development* (pp. 39–57). London: Routledge.

Jagacinski, R. J., & Flach, J. (2003). *Control theory for humans.* Mahwah, NJ: Lawrence Erlbaum Associates.

Jamieson, G. A., & Miller, C. A. (2000). Exploring the "culture of procedures." *Proceedings of the 5th International Conference on Human Interaction with Complex Systems,* pp. 141–145.

Kantowitz, B. H., & Sorkin, R. D. (1983). *Human factors.* New York: Wiley.

Kirlik, A. (1993). Modeling strategic behavior in human-automation interaction: Why an "aid"can (and should) go unused. *Human Factors, 35*(2), 221–242.

Kirlik, A. (1995) Requirements for psychological models to support design: Toward ecological task analysis. In J. Flach, P. Hancock, J. Caird, & K. J. Vicente (Eds.), *Global perspectives on the ecology of human-machine systems* (pp. 68–120). Mahwah, NJ: Lawrence Erlbaum Associates.

Kirlik, A. (1998a). *The ecological expert: Acting to create information to guide action.* Paper presented at the Fourth Symposium on Human Interaction with Complex Systems. Dayton, OH. Retrieved Jan 1, 2004, from http://computer.org/proceedings/hics/8341/83410015abs.htm

Kirlik, A. (1998b). The design of everyday life environments. In W. Bechtel & G. Graham (Eds.), *A companion to cognitive science* (pp. 702–712). Oxford, UK: Blackwell.

Kirlik, A. (2001). On Gibson's review of Brunswik. In K. R. Hammond & T. R. Stewart (Eds.), *The essential Brunswik* (pp. 238–242). New York: Oxford University Press.

Kirlik, A., & Bisantz, A. M. (1999). Cognition in human-machine systems: Experiential and environmental aspects of adaptation. In P. A. Hancock (Ed.), *Handbook of perception and cognition: Human performance and ergonomics* (2nd ed., pp. 47–68). New York: Academic Press.

Kirlik, A., Miller, R. A., & Jagacinski, R. J. (1993). Supervisory control in a dynamic, uncertain environment I: A process model of skilled human-environment interaction. *IEEE Transactions on Systems, Man, and Cybernetics, 23*(4), 929–952.

Kirlik, A., Walker, N., Fisk A. D., & Nagel, K. (1996). Supporting perception in the service of dynamic decision making. *Human Factors, 38*(2), 288–299.

Kirsh, D. (1995). The intelligent use of space. *Artificial Intelligence, 73,* 31–68.

Kirsh, D. (1996). Adapting the environment instead of oneself. *Adaptive Behavior, 4*(3/4), 415–452.

Kirsh, D. (2001). The context of work. *Human-Computer Interaction, 16,* 305–322.

Mandler, G. (2001). Apart from genetics: What makes monozygotic twins similar? *The Journal of Mind and Behavior, 22*(2), 147–160.

McGill, W. J. (1954). Multivariate information transmissions. *Psychometrika, 19*(2), 97–116.

Monk, A. (1998). Cyclic interaction: A unitary approach to intention, action and the environment. *Cognition, 68,* 95–110.

Neisser, U., Boodo, G., Bouchard, T. J., Boykin, A. W., Brody, N., Ceci, S. J. et al. (1996). Intelligence: Knowns and unknowns. *American Psychologist, 51,* 77–101.

Newell, A. (1992). Author's response. *Behavioral and Brain Sciences, 15*(3), 464–492.

Newell, A., & Simon, H. A. (1972). *Human problem solving.* New York: Prentice-Hall.

Norman, D. A. (1988). The psychology of everyday things. New York: Basic Books.

Olson, G. M., & Olson, J. S. (1991). User-centered design of collaboration technology. *Journal of Organizational Computing, 1,* 61–83.

Parasuraman, R., & Riley, V. (1997). Humans and automation: Use, misuse, disuse, abuse. *Human Factors, 39,* 230–253.

Pauls, J. (2001). Life safety standards and guidelines focused on stairways. In W. F. E. Preiser & E. Ostroff (Eds.), *Universal Design Handbook* (pp. 23-1–23-20). New York: McGraw-Hill.

Payne, J. W., & Bettman, J. (2001). Preferential choice and adaptive strategy use. In G. Gigerenzer & R. Selten (Eds.), *Bounded rationality: The adaptive Toolbox* (pp. 123–146). Cambridge, MA: MIT Press.

Polanyi, M. (1966). *The tacit dimension.* New York: Doubleday.

Raab, M., & Gigerenzer, G. (in press). Intelligence as smart heuristics. In R. J. Sternberg & J. E Pretz (Eds.), *Cognition and intelligence.* New York: Cambridge Univeristy Press.

Reder, L. M., & Schunn, C. D. (1999). Bringing together the psychometric and strategy worlds: Predicting adaptivity in a dynamic task. In D. Gopher & A. Koriat (Eds.), *Attention and Performance XVII: Cognitive Regulation of Performance: Interaction of Theory and Application* (pp. 315–342). Cambridge, MA: MIT Press.

Rothrock, L., & Kirlik, A. (2003). Inferring rule-based strategies in dynamic judgment tasks: Toward a noncompensatory formulation of the lens model. *IEEE Transactions on Systems, Man, and Cybernetics—Part A: Systems and Humans, 33*(1), 58–72.

Sanders, M. S., & McCormick, E. J. (1993). *Human factors in engineering and design,* 7th ed. New York: McGraw-Hill.

Schiffer, M. B. (Ed.). (2001). *Anthropological perspectives on technology.* Albuguerque: University of New Mexico Press.

Schmidt, F. L., & Hunter, J. E. (1998). The validity and utility of selection methods in personnel psychology: Practical and theoretical implications of 85 years of research findings. *Psychological Bulletin, 124,* 262–274.

Schooler, C. (2001). The intellectual effects of the demands of the work environment. In R. J. Sternberg & E. L. Grigorenko (Eds.), *Environmental effects on cognitive abilities* (pp. 363–380). Mahwah, NJ: Lawrence Erlbaum Associates.

Sheridan, T. B. (1989). Supervisory control of renote manipulators, vehicles, and dynamic processes: Experiments in command and display aiding. In W. B. Rouse (Ed.), *Advances in Man-Machine Systems Research Vol. 1* (pp. 49–138). Greenwich, CT: JAI Press.

Shiffrin, R. M., & Dumais, S. T. (1981). The development of automatism. In J. R. Anderson (Ed.), *Cognitive skills and their acquisition* (pp. 111–140). Hillsdale, NJ: Lawrence Erlbaum Associates.

Simon, H. A. (1956). Rational choice and the structure of environments. *Psychological Review, 63,* 129–138.

Sternberg, R. J. (1988). *The triarchic mind.* New York: Penguin Books.

Sternberg, R. J. (1997). *Successful intelligence.* New York: Plume.

Sternberg, R. J. (2002). *Intelligence is not just inside the head: The theory of successful intelligence.* In J. Aronson (Ed.), *Improving Academic Achievement* (pp. 228–244). New York: Academic Press.

Sternberg, R. J., & Wagner, R. K. (Eds.). (1994). *Mind in context: Interactionist perspectives on human intelligence.* Cambridge, UK: Cambridge University Press.

Suchman, L. A. (1987). *Plans and situated actions.* New York. Cambridge University Press.

Suchman, L. A. (2000). Embodied practices of engineering work. *Mind, Culture, and Activity,* 7(1&2), 4–18.

Sunstein, C. R. (Ed.). (2000). *Behavioral law & economics.* New York: Cambridge University Press.

Tolman, E. C., & Brunswik, E. (1935). The organism and the causal texture of the environment. *Psychological Review, 42,* 43–77.

van Geert, P. (2003). Measuring intelligence in a dynamic systems and contextualist framework. In R. J. Sternberg, J. Lautrey, & T. L. Lubart (Eds.), *Models of intelligence: International perspectives* (pp. 195–212). Washington, DC: American Psychological Association.

Vicente, K. J. (1999). *Cognitive work analysis.* Mahwah: NJ: Lawrence Erlbaum Associates.

Vygotsky, L. S. (1981). The problem of the cultural development of the child, II. The instrumental method in psychology. In J. V. Wertsh (Ed.), *The concept of activity in Soviet psychology* (pp. 134–143). Armonk, NY: M. E. Sharpe. (Reprinted from *Journal of Genetic Psychology, 36,* 414–434, 1929)

Walker, N., & Fisk, A. D. (1995, July). Human factors goes to the gridiron. *Ergonomics in Design,* 8–13.

Warren, W. H. (1984). Perceiving affordances: Visual guidance of stair climbing. *Journal of Experimental Psychology: Human Perception and Performance, 10,* 683–703.

Wickens, C. D., Gordon, S. E., & Liu, Y. (1998). *An introduction to human factors.* New York: Addison-Wesley.

Woods, D. D. (2003). Discovering how distributed cognitive systems work. In E. Hollnagel (Ed.), *Handbook of cognitive task design* (pp. 37–53). Mahwah, NJ: Lawrence Erlbaum Associates.

Zhang, J., & Norman, D. N. (1997). The nature of external representations in problem solving. *Cognitive Science, 21*(2), 179–217.

7

Cooperation Between Human Cognition and Technology in Dynamic Situations

Jean-Michel Hoc
Centre National de la Recherche Scientifique
Institut de Recherche en Communications et Cybernétique de Nantes
University of Nantes

Clearly, the study of the relationships between humans and technology is justified because humans are becoming increasingly immersed in technological environments—at work, at leisure, or in everyday life. Technology provides humans with external support to cognition, not only representations, but also "intelligent" agents contributing to the execution of human tasks. Several authors have stressed the role of external objects and agents that are "integrated" into human cognition (Klahr, 1978; Zhang & Norman, 1994). Sometimes, with the development of automation, the machines are not actually designed to fulfill the function of assisting the humans, but to act as partners. This results in two properties of such technical environments. First, humans can only exert a partial control because the machines can have a minimal autonomy. Second, humans must, properly speaking, cooperate with machines—the tasks being executed by a combination of humans' and machines' actions. Thus, it is not reasonable to exclude technological environments from the study of human cognition. Cognitive psychology elaborates knowledge that is relative to social and technological contexts, because cognition is determined both by human "nature" and human "culture." If scientific psychologists had been around at the time of our prehistoric ancestors, they would possibly have stressed different aspects of cognition than they would today. Certainly, the study of the cognitive properties of human–computer interfaces would not have been considered in prehistoric research.

Beyond the intrinsic value of the study of the relationships between humans and technology, there is also scientific interest in this kind of study. The confrontation of technology is sometimes a useful method for gaining access to hidden aspects of human cognition. When a computer is utilized as a support to representation or as a medium to perform cognitive tasks, covert aspects of human cognition can become easily accessible. For example, computer programming is an informative situation to study human planning because of the use of explicit codes to express plans (Hoc, 1988). However, a computer does not operate like humans and difficulties encountered by humans when using or programming computers can be a means of accessing human strategies in a negative way.

This chapter is devoted to the study of human cognition in dynamic environments, such as air traffic control, aircraft piloting, industrial process control, and car driving. In these situations, the human operators exert partial control on events. Many situations of this kind imply high-level technology, that is to say, autonomous machines capable of some intelligent behavior. A variety of definitions of intelligence are proposed in this volume, each one stressing a particular aspect of intelligence. In this chapter, I will stress the adaptive power of intelligence. Thus, intelligence will be defined as the capability of a cognitive system to adapt to a certain set of circumstances. We will see that adaptation has two faces. On the one hand, the system applies its knowledge to assimilate a situation to a well-known one. On the other hand, it modifies its knowledge to resolve serious mismatches between the two when its objectives are jeopardized.

After presenting the main (cognitive) features of dynamic situations, two main kinds of research results within this context will be stressed. First, the study of dynamic situations, because of their partial control and uncertainty, is an appropriate way to gain access to human adaptation mechanisms and cognitive control modalities. Emphasis will be on the concept of situation mastery, which is the main motivation of adaptation. Second, the presence of autonomous machines poses very clearly the problems associated with human–machine cooperation. This latter concept will be approached with theoretical and methodological tools currently in use within the study of human–human cooperation, with some restrictions.

DYNAMIC SITUATIONS

Dynamic situations, considered as partially controlled by the human subject or operator (the latter being more appropriate in this context), occur frequently in everyday life (e.g., walking through the crowd on a sidewalk), in leisure (e.g., playing a game or sport with an opponent), or work (e.g., industrial process control). In the study of work, several cognitive features of this kind of situation have been stressed—partial control, temporal dynamics, multiple representation and processing systems, uncertainty and risk, and time-sharing between several tasks.

Partial Control

A dynamic situation is, by definition, partially controlled by the human operator, as opposed to a static situation, where nothing can happen without human intervention. As noted by Bainbridge (1988), this implies that human operators use two kinds of knowledge—knowledge of the (technical) process under supervision and knowledge of their goals. In static situations where everything is determined by the operators' goals and actions, at least for experts, *egocentric* knowledge of their own goals or actions is sufficient to perform tasks because it is these that fully determine the changes in the environment.

For example, a keystroke on the computer keyboard fully determines what will happen to the text being written. That is why several models of human–computer interaction have been elaborated on the basis of keystroke analysis (Card, Moran, & Newell, 1983, and others). However, the situation is quite different in, for example, ship navigation. When the watch officer acts (e.g., to adjust the helm angle), this will enter into the determination of the future trajectory, but so too will other factors, such as wind, current, and inertia (response delay). Cooperation is one of the dynamic features of situations. For example, at sea, collision avoidance leads to the management of uncertainty over other ships' maneuver intentions, sometimes forcing these intentions in order to increase the control of the situation, at the price of not applying the international regulation (Andro, Chauvin, & Le Bouar, 2003). Because of the intervention of factors that are not always fully predictable, a large part of the human operator's activity is devoted to diagnosis in order to understand the past, present, and future trend of the technical process. Several studies on blast furnace control have stressed the role of diagnosis in the design of an expert system that advises operators on diagnosis or action (Hoc, 1989). The expert system's strategy was similar to the operators' strategy that decomposed the process into seven covert functions (e.g., internal thermal state, reduction quality resulting in the transformation of iron oxide into iron, etc.), and evaluated each one on the basis of overt parameters (e.g., cast iron temperature for thermal state) that played the role of symptoms in order to access syndromes (covert functions). Obviously, the anticipated effects of the operators' actions were considered when developing diagnosis. The prominence of diagnosis has also been stressed in more proceduralized situations, such as nuclear power plant supervision, where operators, while applying standard procedures, continue diagnosis in order to be sure that the situation evolution is still compatible with the validity conditions of the procedures (Roth, 1997).

The partial control of dynamic situations has introduced the concept of *situation awareness* (Endsley, 1995). This concept stresses the fact that the human operators must regularly update their representations of the technical processes. However, as suggested, the human operator must be introduced into the definition of a situation, which should be considered as the interaction between a task and a human operator (Hoc, 1988). For example, car driving is not the same situation for an ordinary driver, for a racing driver, or for a mechanic, because they do not share the same declarative and procedural knowledge, which results in very different task definitions. Dynamic situation management does not only imply the maintenance of an adequate "picture" of the external process dynamics, but also the elaboration and use of metaknowledge on one's resources (declarative and procedural knowledge, as well as available energy in terms of workload or motivation). The main objective is to ensure that there is consistency between the demands of the task and the operator's resources. For example, air traffic controllers do not, properly speaking, resolve trajectory conflicts between aircraft, but try to avoid problems in the future, managing their workload as well as the air traffic.

Temporal Dynamics

Another important objective in dynamic situations is the synchronization of technical process dynamics and cognitive process dynamics. Certainly, a desynchronization is often the result of a lack of consistency between the demands and the resources. It can also be produced by bias in temporal estimation (De Keyser, 1995). Several studies have been conducted in which the technical process speed has been varied (e.g., Hoc, Amalberti, & Plee, 2000). They show the need for a multiprocessor model of the human operator in dynamic situations.

Technical process speed can be considered as the technical process's bandwidth, that is to say, the minimal frequency of information gathering needed in order to avoid missing a crucial event for which a response is required. Indeed, response delay must be considered. When information on an event occurs too late in relation to response delay, the technical process is not controllable (for example, when having a shower it is very difficult to control the water temperature if someone else extracts water randomly).

In studies, it was shown that the increase in process speed resulted in a parallelism between cognitive control modalities. The impossibility of planning in real time, on the basis of symbolic processes, led to an initial planning followed by a subsymbolic control (on the basis of signals rather than signs), with the intention to minimize parallelism with replanning activities, at the symbolic level and without immediate effects. Fighter aircraft mission planning and execution is a typical example of this strategy (Amalberti & Deblon, 1992), confirmed in an experiment comparing varying process speed in a firefighting microworld (Hoc et al., 2000).

Multiple Representation and Processing Systems (RPS)

Because of the partial control of dynamic situations, it is impossible to confine them to state transformation situations, widely studied by the cognitive psychology of problem solving. The technical process cannot be controlled only by a transformational RPS (Hoc, 1988), which supports transitions between states by operations to be applied by the human operators. Types of interventions other than those of the human operators are possible, and the conception must be more exocentric. Three other types of RPS have been widely developed in this kind of situation. In blast furnace control, the RPS was mainly functional (related to the system goals) and causal (related to physical laws), in order to support diagnosis/prognosis and action decision at the same time (Hoc & Samurçay, 1992). In certain cases, a topographic RPS is utilized in order to match diagnosis and the relevant part of the plant (Rasmussen, 1986).

Uncertainty and Risk

Prognosis for action decision is difficult when there is uncertainty about the course of events. An element of parallelism is introduced into the cognitive process. At the same time, the human operator favors the most likely result while remaining prepared to deal with any unexpected situations. Contingent planning has been shown to be a solution to this problem (Amalberti & Deblon, 1992). However, there are associated costs. In addition, expert operators may have metaknowledge at their disposal, which convinces them that they would be able to manage the unexpected situation in real time by triggering routines. The problem would be simple if it could be reduced to uncertainty. However, dynamic situations, especially industrial ones, open the way to high-level costs associated with bad consequences of events (e.g., nuclear power plant control, fighter aircraft piloting, etc.), either on the environment (e.g., radioactive pollution) or on the operators themselves (e.g., missile threat). Risk management is related to the processing of ratios between uncertainties and costs. A risk is the consequence of an event (associated with a probability and a cost) that the operators do not consider in their plans because of cognitive cost or motivation. A high level of uncertainty in an event where consequences are very costly will lead to the risk being considered seriously. On the other hand, a low uncertainty in an event where consequences are not very serious will lead to the risk being disregarded. For example, airline pilots commit many errors, even when they are experts, but these errors have no serious consequences and their recovery would be costly (Amalberti, 2001). More often than not, operators evaluate costs in terms of their own activities. The consequence most often considered is the loss of situation control. Thus, situation mastery (see the following) is a key issue within the context of risk management.

Time-Sharing Between Several Tasks

Although many experimental studies confront subjects with mono-task activities for reason of reduction and clarity of results, in actual work settings operators are very seldom doing one thing at a time. Some dynamic situations are typical of this task parallelism, for example, in air traffic control (aircraft entries add new problems, sometimes more urgent, to current problems), and medical telephone emergency services (incoming calls can also create new problems, more urgent than the current problems). Task concurrence is not under human control, even if some adjustments are possible, for example, when delaying an aircraft entry in air traffic control. One of the main activities for a human operator in this kind of situation is avoiding too much of a concurrence between tasks (Morineau, Hoc, & Denecker, 2003).

ADAPTATION AND COGNITIVE CONTROL

Our conception of intelligence is very close to that of Piaget (1974), who linked intelligence to adaptation as an "equilibration" between two complementary mechanisms—assimilation and accommodation (following the biological metaphor). Cognitive control is the means to implement adaptation (and intelligence).

Adaptation and Situation Mastery

Classically, adaptation is considered to be an adjustment to environmental conditions, as if the mechanism, in the context of cognition, consists in modifying a knowledge structure to fit in to environmental conditions. As a matter of fact, the case is more complex than it appears at first glance. Piaget (1974) defined adaptation as the product of a process of "equilibration" (search for an acceptable balance) between assimilation and accommodation. This approach is very close to that adopted by Lazarus (1966) when describing coping strategies adopted to reduce stress. In work situations, but also in everyday life, human operators are confronted with two opposite requirements. They face task demands, in terms of performance quality, and at the same time resource needs (declarative and procedural knowledge, and energy—mental workload and motivation) in order to satisfy these demands. Following on from Simon (1983), we can think that bounded rationality is the reason for the search for an acceptable balance between task demands and internal resources (or a cognitive compromises; Amalberti, 1996). The question of external resources is not addressed here, but it is also an important one to consider, especially when cognition uses external resources as well as internal representations (Zhang & Norman, 1994). The search for an acceptable balance is determined by the feeling of situation mastery.

In dynamic situations, the human operators' main objective is to maintain the situation within certain limits where they know they can keep it under an acceptable degree of control. The classical notion of optimal performance is replaced by a notion of satisfactory or acceptable performance, in relation to social or other pressures. The operators are not ready to invest their entire resources in order to keep back a minimal amount to face unexpected and demanding situations. They can accept a certain cost, but not that much. The reason for these limitations is mainly attributed to bounded rationality. Operators know that they cannot consider every possible contingency without taking the risk of exceeding their resources. So the main risk they manage is that of losing situation mastery. This raises several questions. First, a gap may exist between the subjective representation of situation mastery and its objective evaluation. No side taken alone is sufficient. The subjective approach enables the observer to understand how the operators manage the cognitive compromise, but it is not sufficient to evaluate the accuracy of the compromise. The objective approach provides the observer with evaluation tools, but only partly, because knowledge of the operators' cognitive resources is necessary. Second, measures taken to ensure short-term mastery can jeopardize medium- and long-term mastery, and vice versa. Thus, a balance must be managed to reach an acceptable mastery from the short to the long term. For example, in air traffic control, turning a particular aircraft to avoid a conflict with another can be an efficient way to solve the short-term problem, but can create a more serious problem in the future with other aircraft in the sector.

Now, we will turn to the two complementary mechanisms described by Piaget (1974) to perform this equilibration—assimilation and accommodation.

Assimilation is the (top-down) process by which already available resources are utilized to manage the situation. From the operators' internal viewpoint, assimilation can succeed in two ways. First, the representation of the situation can be simplified to fit into an already known situation. This procedure can be successful if this abstraction drops situational features that are not necessary for reaching the action objective. Second, operators can act so that the situation does not deviate from a well-known or controlled situation. This strategy has been described not only in relation to fighter aircraft mission preparation (Amalberti & Deblon, 1992), but also in collision avoidance at sea where some maneuvers, not covered by the regulations, can be interpreted as taking control over the situation (Andro et al., 2003). The common feature of these strategies is to act in such a way that the situation could never develop outside the envelope of resources. Very often, it bears on a hierarchy of constraints in the situation and can be described as constraints relaxation. Assimilation is one of the most powerful mechanisms for adaptation at the lowest cost.

However, following Piaget, if there is no "resistance" from the reality, pure assimilation can lead to the opposite of adaptation. When errors are committed, when the objective is not reached, information feedback is utilized in order to devote more resources to the task. One possibility is that learning results in new resource elaboration. Accommodation takes place when assimilation does not fully succeed in reaching an acceptable level of performance. It represents a certain cost and is brought into action by operators either when the current cognitive compromise opens up the risk of losing situation mastery or when available resources are sufficient to use accommodation. A series of experiments undertaken with the NEWFIRE microworld (Løvborg & Brehmer, 1991) provides a clear illustration of the limits of this propensity to accommodate beyond the operators' capabilities (Hoc et al., 2000). A main factor in these experiments was process speed. When speed was reduced, this had a positive effect at first. After this, performance deteriorated because, at the same time, the subjects thought that the risk of losing situation mastery was low and they tried to improve the quality of their performance. However, their model of the technical process was not sufficient to enable them to refine their strategy at this high degree of detail, and this attention to the details caused their high-level control to deteriorate.

Cognitive control and, more precisely, the parallelism between several cognitive control modalities is a means to succeed in this adaptation leading to situation mastery.

Cognitive Control Dimensions

Cognitive control can be considered as the instance that enables the cognitive mechanisms and representations (symbolic or otherwise) to be brought into play with the appropriate temporal and intensity features in order to realize adaptation. Here, the approach to cognitive control is shared with that of Amalberti (Amalberti & Hoc, 2003), with whom I have closely collaborated on this topic for a long time. This approach springs from two relevant theoretical accounts. The first one—the hierarchy of cognitive control levels proposed by Rasmussen (1986)—is straightforward because the theory is taken without new interpretation, only with some further developments and some changes in the terminology. The second one—the contextual control modalities introduced by Hollnagel (1993)—will be used in a quite different sense. Reference to this theory is a psychological reinterpretation of a phenomenological approach. These two theoretical contributions suggest two orthogonal dimensions to categorize the cognitive control modalities. In this chapter, I have chosen to stress the data (information) necessary for the control rather than the control mechanism in itself.

Control Data Abstraction Level. The data necessary for the control can be symbolic or subsymbolic. The distinction here is between data that need to be interpreted before use and data that directly trigger a response.

Within the framework introduced by Rasmussen, two kinds of symbolic data are defined: signs and concepts. A sign has two faces, form and content (e.g., the indication "water in diesel oil" on the dashboard). The form is directly perceptible (the icon), but the access to the content (the fact that there is water in the oil) needs an interpretation (sometimes a glance at the manual). Signs are utilized by rule-based behaviors (when procedure execution is not automatic, but guided by symbolic attentional control). A concept is more complex because it integrates a knowledge structure, sometimes very rich, and some procedural knowledge associated with the structure. For example, the concept of volume groups together the tridimensionality of the concept and calculus formula. Concepts are implied in knowledge-based behaviors (problem-solving situations where the procedure is not straightforward).

Sometimes, the behavior is in reaction to (subsymbolic) signals without symbolic interpretation. For example, an expert driver stops when the traffic light turns to red without reconstructing the content "stop" behind the form "red." The stimulus is not processed as a sign, but as a signal. This process, involving the generation of signals from signs, is not unique. A reference to Gibson's theory of affordances would be useful here (Gibson, 1986). The notion of affordance assumes that what the subjects perceive is the product of an interaction between the subjects' needs or objectives and the objects' properties. In other words, perception directly triggers the relevant properties of objects in relation to the context (action objectives). Signals are affordances. Gibson's theory must be enlarged to integrate signals originating from signs. In this case, the affordance is expert, in the sense that a certain amount of expertise (learning by doing) is necessary to have the affordance available as such. Some stimuli present affordances without needing a specific expertise. They correspond to basic affordances, like the reaction provoked by a fire perceived as a threat in certain circumstances (e.g., in a forest), or the attraction in other ones (e.g., next to a fireplace in winter).

More often than not, this distinction between symbolic and subsymbolic cognitive control is related to attention and automaticity. Certainly, symbolic control needs absolute attention and corresponds to serial processes rapidly reaching the mental workload limit. Correctly speaking, one should specify that this type of attention is of a symbolic kind. There also exists subsymbolic attention, for example, visual attention. Subsymbolic processes are always automatic. Thus, roughly speaking, confusion between the classical controlled/automatic opposition and the symbolic/subsymbolic opposition is possible. However, an automatic process is also controlled, but not directly by symbolic representations.

Control Data Source

On the basis of the structure of behavior, Hollnagel (1993) proposed another dimension to classify the cognitive control modalities in his contextual control model, COCOM. The main difference between this and Rasmussen's model is that it is phenomenological rather than psychological. A psychological interpretation of Hollnagel's model is possible, although it is not exactly equivalent. Hollnagel's intention was not to elaborate a model that was useful for psychologists, but a model that was workable by system designers, on the sole basis of the observation of behavior. The dimension introduced by Hollnagel is comprised of four values—scrambled control, opportunistic control, tactical control, and strategic control. Basically, Hollnagel presents this dimension as representing the temporal span covered by the control.

The control is scrambled when the subject's behavior appears to be random. Several psychological interpretations are possible. One of them could be that the subject is reacting to environmental affordances that cannot shape behavior within a clean structure. Thus, behavior is determined by an unstructured series of stimuli. It is not random in the sense of an intrinsically random property, but in the sense of a causal determination by quite random stimuli. Many car accidents can be described as an environment that has trapped the driver into a series of inappropriate affordances (e.g., turning the wheel to the right on ice when the car slips to the left). The control is opportunistic when it presents certain logic without deep planning. Clearly, the subject can be either directed by structured affordances that result in highly structured behavior, or directed by more expert affordances, triggering a series of actions instead of isolated actions, and thus covering a larger temporal span (e.g., turning the wheel at first to the left when the car slips to the left on ice, in order to recover adhesion, before turning the wheel to the right). The two other control modalities are more difficult to interpret psychologically, except that they can be sorted by increasing temporal scan width. The control can be tactical or strategic. The boundary between these last two values is probably more determined by the domain than by a psychological theory. However, the last two control modalities result in a more planned behavior than the first two.

In many cases, this distinction between cognitive control modalities in terms of temporal span can be seen as a result of two other highly correlated dimensions. The first one is related to anticipation and enables us to separate reactive (or closed loop, or by feedback) control and anticipative (or open loop, or feedforward) control. Scrambled control and opportunistic control are reactive, whereas tactical control and strategic control are anticipative. A second dimension is also correlated with temporal span and with anticipation. I have retained its formulation for the cognitive control approach, with the other formulations understood (temporal span and reactive/anticipative control). It is related to the source of the data utilized for the control. The control can be external, bearing mainly on data found in the environment (a kind of data-driven process), or internal, bearing mainly on internal (symbolic or not) representations (a kind of knowledge-driven process). In the first case, the control is reactive and with a restricted temporal span. In the second case, the control is anticipative and has a longer temporal span than the first one.

Cognitive Control Modalities

With both these dimensions, the model summarizes a more complex picture in which other dimensions could be brought into consideration, for example, the distinction between attentional and nonattentional processes. However, such a restriction enables us to examine many questions of interest. Are the two dimensions orthogonal or not? I think they are. Can behavior be governed by diverse control modalities at the same time? I think that the control modalities can act in parallel, or at least in a time-sharing way that is very close to the idea of a parallelism. A third question concerns the relationships between modalities when they act in parallel. A fourth question is related to the transitions from one modality to another.

It is easy to find examples for each of the four modalities generated by the crossing of the two dimensions.

- Symbolic and internal control. The elaboration of a strategic plan covering a long temporal span.
- Symbolic and external control. The assistance of a procedural interface guiding the activity step by step.
- Subsymbolic and internal control. The execution of well-learned and complex routines as is the case of musical execution by heart, guided by expert affordances.
- Subsymbolic and external control. The erratic guidance of basic affordances taken from within the environment.

Many studies show that these modalities can act in parallel. That is why it is difficult to identify their occurrence. A number of studies show that symbolic processes are often utilized to supervise subsymbolic processes. Conversely, subsymbolic processes can send information to symbolic process by emergence. These two relationships between the control modalities are crucial in risk management, particularly in error management. The transitions between the control modalities are more complex than can appear at first glance. Globally, expertise is likely to transform symbolic processes into subsymbolic ones, which are less costly than the symbolic ones. However, minimal symbolic processes are brought into play to supervise the subsymbolic processes in order to manage errors (error prevention, consequence mitigation, etc.).

COGNITIVE COOPERATION

The presence of autonomous machines in dynamic situations where humans are also acting introduces the need for applying the theoretical framework of cooperation to the study of human–machine relationships. On the one hand, the know-how of machines (their knowledge in their domains of intervention) has considerably increased over recent decades, resulting in an increase in the intelligence of machines in terms of adaptation power. On the other hand, the ability of machines to cooperate (their knowledge in terms of cooperation, especially with humans, or their "know-how-to-cooperate") remains very restricted. Certainly, a number of studies have been devoted to the cooperative capabilities of machines, but these have mainly been conducted in environments where time constraints are not as onerous as in dynamic situations. That is why I have restricted the framework to a narrow approach of cooperation, stressing its cognitive aspects and neglecting emotional or social aspects (Hoc, 2001). This approach borrows concepts from the study of human–human cooperation in order to apply them in the domain of human–machine cooperation. I proposed the following definition of cooperation, grouping together its minimal properties:

Two agents are in a cooperative situation if they meet two minimal conditions:

1. Each one strives toward goals and can *interfere* with the other on goals, resources, procedures, etc.
2. Each one tries to manage the interference to *facilitate* the individual activities and/or the common task when it exists. (Hoc, 2001, p. 515)

The definition and the studies conducted lead toward an analysis of the cooperative activities rather than the cooperative structures (organization of the network between the agents). Let's consider the two main aspects of the definition—interference and facilitation.

Interference and Facilitation

The notion of interference is well known in physics, where it does not have the negative connotations implied by the popular use of the term. Two signals interfere if they reduce or reinforce each other. The notion is also very well known in artificial intelligence, especially in the context of planning studies where goals are not always independent and where reaching one goal can jeopardize the attainment of another. Castelfranchi (1998) proposed that the notion of interference be applied to cooperation, meaning that "the effects of the action of one agent are relevant for the goals of another: i.e., they favor the achievement or maintenance of some goals of the other (positive interference), or threaten some of them (negative interference)" (p. 162). Cooperative activities are those that are implied by interference resolution in real time. Thus, we exclude situations where such a resolution is executed beforehand and in such a way that the agents can act independently of each other. In studies, I have identified four types of interference, although this list is not exhaustive. (a) Precondition interference is created by the need for an agent to have another agent performing some activity before triggering its own. (b) Interaction interference corresponds to the combination of precondition interference in two directions. (c) Mutual control interference is the opportunity for an agent to check the activity of another and to detect errors or suboptimality. (d) Redundancy interference is the necessary condition for opening the way to function allocation. When several agents are able to ensure a function, the strength of the multiagent system lies in its capability to adapt to situations (e.g., replacing one unavailable or misplaced agent by another), but at the additional cost of diagnosing the problem and deciding the allocation.

The only difference between cooperation and competition is that the former aims at facilitating the others' activities, whereas the latter aims at being an impediment to them. Such facilitation is a difficult notion because it is not necessarily symmetric. If the function performed by one agent has priority, the other agents' cooperative activities may aim at facilitating this particular agent's activity at the risk of rendering their tasks more difficult. It is the same for the facilitation of a common goal, which can result in overloading individual activities. However, cooperation does not mean a common task. Each agent can have very distinct tasks but still interfere on resources (e.g., sharing a printer in an office).

Cooperative activities can be classified according to a dimension that mixes the temporal span of their results and their abstraction level—action, planning, and meta level.

Action Level

At the action level, the cooperative activities aim at resolving short-term problems by local interference creation, detection, resolution, and anticipation. As a subproduct, they can also contribute to maintain a COmmon Frame Of Reference (COFOR; see the following) between the agents. Interference creation is noted when it is voluntary and consists of mutual control between the agents. Interference anticipation is related to knowledge in the domain that enables an agent to infer the goal of another on the sole basis of the observation of its behavior.

Planning Level

At the planning level, a COFOR is elaborated and maintained, facilitating the performance of the action level cooperative activities in the medium term. Each agent manages a private current representation of the situation, integrating the environment and the agent's internal state. The concept is close to situation awareness, but with the integration of the latter. Situation awareness only concerns the environment. This conception of a situation stresses the interaction between an agent and a task so that the current representation of the situation is not restricted to the environment.

When several agents are cooperating, they each have distinct current representations of the situation. There are, however, some relationships between these current representations. Roughly speaking, one can say that they share a common intersection that corresponds to the COFOR. However, most of the time, each agent has a distinct COFOR, that is to say, distinct representation. The notion of COFOR is more one of compatibility than of identity. What is in common is an abstract representational system, but the implementation of this system by each agent is different and depends on the agent's goals and viewpoint on the tasks. Thus, to be more precise, each agent has an individual COFOR that is part of the individual representation of the situation. Such a COFOR can be comprised of certain representations that the individual assumes to be shared by the other agents, whereas they are not.

The COFOR concerns the environment (in dynamic situations, the controlled process) and the team's activity (the control activity). Among others, the latter includes common goals, common plans, and function allocation (function distribution among the agents). Part of COFOR maintenance is performed by explicit communications, but another part is obtained implicitly by the execution of action level cooperative activities.

Meta Level

The last abstraction level covers a much larger temporal span. This meta level is elaborated after a certain experience of cooperation between the agents. It mainly concerns the elaboration of a common communication code, of compatible representations, and of models of oneself and of the partners.

COMMON FRAME OF REFERENCE (COFOR)

Studies in dynamic situations have shown the importance of the elaboration and maintenance of a COFOR between the agents. Two kinds of situations have been studied—air traffic control and fighter aircraft piloting. In air traffic control (in France), two controllers must cooperate to manage a sector. A radar controller is in charge of safety (aircraft conflict resolution) and expedition (mainly timetabling) of the aircraft in the sector. A planning controller manages the intersector coordination and assists the radar controller. In addition to the human–human cooperation, I have also studied human–machine cooperation between the radar controller and an automatic conflict resolution device. In order to import human–human cooperation features into the design of the automatic device, I have studied human–human cooperation in an artificial situation where the aircraft were distributed between two radar controllers, so that they were forced to cooperate over conflict resolution (Hoc & Carlier, 2002). Fighter aircraft piloting has been studied in two-seater aircraft managed by a pilot and a weapon system officer (Loiselet & Hoc, 1999). The pilot mainly takes care of short-term activities (aircraft piloting and firing). The weapon system officer performs the navigation tasks and prepares the firing tasks. However, some overlapping between the two kinds of activity and some function allocation between the two roles are possible. Cooperation in planning (COFOR elaboration and maintenance) occurs much more frequently in air traffic control (80%)

than in fighter aircraft piloting (50%). The process is slower in the first case than in the second case, where mission preparation is necessary to establish a COFOR, impossible to entirely elaborate in real time.

In the two domains COFOR elaboration and maintenance have been distinguished on the basis of the number of speech turns. Several turns correspond to some negotiation between the agents that disagree on the analysis of the situation. They are related to COFOR elaboration aiming at reaching a consensus. One or two turns, the second one being an acknowledgment, correspond to COFOR maintenance. The receiver agrees and has only to integrate the new information into its COFOR. In fighter aircraft piloting, 80% of the planning-level cooperative activities were devoted to COFOR maintenance (as opposed to elaboration). In air traffic control, the figure was 65%. With a difference because of time constraints, in both cases, COFOR maintenance occurs much more frequently than COFOR elaboration. In air traffic control, action-level cooperative activities were mainly devoted to mutual control and interference anticipation. The implication for human–machine cooperation design may be quite optimistic. It is difficult for a machine to manage the elaboration of a COFOR under time constraints, but it is easy to have it participating in COFOR maintenance.

These results are consistent with other studies showing the importance of COFOR in cooperation. Entin and Serfaty (1999) found that cooperative work is improved when the team managers communicate short summaries of their current representations of the situation. Other authors have shown the benefits of a continuous updating of the COFOR (Heath & Luff, 1994; Paterson Watts-Perotti, & Woods, 1999).

COFOR Structure

In both situations, COFOR management mainly concerns the control activity (64% in air traffic control; 53% in fighter aircraft piloting, as opposed to the process under control). This result justifies our caution when using the notion of situation awareness, restricted to a representation of the environment. COFOR is also composed of representations of the team activity and they must be communicated to the agents.

Identical or Compatible Representation

A recent study in fighter aircraft piloting (Loiselet, 2002; Pacaux-Lemoine & Loiselet, 2002) showed that the COFOR should be seen as being composed of compatible rather than identical representations. In this situation, the human operators had the choice to either display identical external support on their Video Display Terminals (VDTs) in the cockpit or to display specific supports to their individual activity. Identical support is very seldom displayed, whereas individual supports are displayed frequently. Because action-oriented external representations are favored over identical representations, the idea of a unique representational system to externalize the COFOR is not relevant when the operators have very different tasks.

FUNCTION DELEGATION

In dynamic situations, the need for the human–machine system to adapt to unexpected situations is not compatible with the constraint of allocating functions beforehand (McCarthy, Fallon, & Bannon, 2000). Allocation must be dynamic and must consider local as well as general conditions. There have been several studies on dynamic allocation in the air traffic control domain.

Dynamic Task Allocation

I have studied the best conditions for dynamic task allocation between radar controllers and an automatic conflict resolution device. Previous studies showed that an explicit allocation (decided by the radar controllers) was less efficient than an implicit one (decided by the machine on the basis of an evaluation of the radar controllers' workload), but was preferred by the controllers (Vanderhaegen, Crevits, Debernard, & Millot, 1994). The improvement of support given to the controllers in the explicit allocation led to acceptable results in terms of an increase in anticipative strategies or in human–human cooperation quality (Hoc & Lemoine, 1998). In the best allocation condition in terms of performance (explicit assisted allocation), the programming controllers were in charge of the allocation, so that the radar controllers' workload was alleviated while still ensuring that they had the right to veto decisions made by the planning controllers. However, two problems remained that led to reconsideration of the notion of task allocation. First, the radar controllers refused a number of task allocations to the machine on the basis of a disagreement over problem definition. As a matter of fact, the automatic device only considered two-aircraft problems, whereas the controllers had also defined three- or four-aircraft problems, sometimes even more. For a human controller, a conflict is comprised of several focal aircraft and several contextual aircraft. When controllers see two (or more) conflicting aircraft (future crossing in less than the minimal separation of distance or time), they are not already sure that it is the problem. They have to envisage a possible plan to resolve the conflict before they can be sure that there are no contextual aircraft, that is to say any aircraft that will constrain the resolution, in order to avoid creating a new conflict while resolving the present conflict between the focal aircraft. Second, when the controllers were not in charge of task allocation, they did not apply a mutual control to the machine (the well-known complacency phenomenon). This way of operating resulted in splitting the supervision field into two impermeable parts.

Dynamic Function Delegation

The limitation of the task allocation notion led us to favor a notion of function allocation (Hoc & Debernard, 2002). Air traffic control is a typical example of a situation where tasks must be defined in real time. Aircraft conflicts appear progressively and are not planned in advance. On the contrary, their existence is evidence of imperfect planning. A central planning instance does exist (e.g., in Europe) that is aimed at reducing the occurrences of conflict, but partially, because of unpredictable events. Even a minor aircraft delay can result in an unexpected conflict. In the case of cooperation between several agents, task definition must belong to the COFOR. If task definition depends on the way it will be done and on a negotiation between the agents, it is clear that the current restricted level of a machine's intelligence will not enable the designers to program it in order to enter into this kind of negotiation. Thus, we have explored the principle of function delegation that, on the basis of an experiment in progress, looks like it could be more acceptable than task allocation.

A function is defined more generically than a task because a function will appear in very different tasks. In the new platform, controllers define the tasks and communicate with the automatic conflict resolution device within the framework of their problem space (regularly updated). Function delegation consists of transferring a problem (focal and contextual aircraft) and a plan to the machine. A plan is a schematic resolution procedure (e.g., having aircraft A turning to the left, going behind aircraft B, etc.). First, the machine will compute an acceptable route that is compatible with the plan. If there is no acceptable route (e.g., because contextual aircraft are discovered that prevent the plan from being adopted), the machine returns some kind of error message indicating that the problem representation is possibly not correct. Second, if there is an acceptable implementation of the plan, the machine will do it and re-route the aircraft after they have crossed. At each step, a confirmation of the delegation will be required from the controller. In this way, task definition is kept within the controllers' control and what is delegated is only part of a task, thus encouraging the controllers to supervise the machine's operation.

FUTURE DIRECTION: COOPERATION AT THE SUBSYMBOLIC LEVEL AND COOPERATION MODES

Future study will continue to elaborate on the topics that have been discussed here, especially the problem of the external support to the COFOR (in human–human cooperation as well as in human–machine cooperation) and the question of function delegation. However, air traffic control and fighter aircraft piloting remain highly symbolic tasks, although they have undeniable subsymbolic components. Verbal reports are at the core of these activities. Air traffic controllers are used to speaking spontaneously to each other, and it is an important part of their activities. Communication with machines in this context is also of a verbal nature. In the cockpit, spontaneous verbal reports are common. In this kind of situation, the study of cooperation is easy because, for the most part, the cooperative activities can be inferred from verbal communications. However, in dynamic situations with a high temporal pressure, human–human cooperation and human–machine cooperation cannot be dealt with by symbolic processes alone.

Automatic devices are being developed that do not interact with humans in a symbolic way. A good example is the increasing development of automatic devices to improve car-driving safety. Adaptive Cruise Control (ACC) and Electronic Stability Program (ESP) are the most widespread devices, the former regulating distances between vehicles, the latter seeking to avoid spinning around (Stanton & Young, 1998). Several research programs are devoted to this kind of device, aimed at contributing to a reduction in car fatalities (e.g., the ARCOS program in France; Hoc & Blosseville, 2003). This type of cooperation leads researchers to find counterparts of interference anticipation, COFOR elaboration and maintenance, function allocation, models of the partners, and so on, at the subsymbolic level. In order to be efficient, information exchange between the devices and the drivers must be sensorial, that is to say restricted to signals that are rapidly acted on, without deep interpretation.

Within the context of car driving, the aim is to keep the drivers as the main process controllers and to avoid transforming them into car supervisors (in the way that pilots have become aircraft supervisors). In relation to this aim, several cooperation modes can be sorted following an increasing intrusion in the drivers' activity.

A *perception mode* is restricted to the presentation of raw information to the driver. The presentation format depends on what is expected from the driver. The information can be symbolic (e.g., speed value) if a symbolic process is required (e.g., to compare the current speed to local enforcement). In this case, signs are processed—forms are perceived and interpreted in terms of content to be useful. The information can be subsymbolic and enter as such into the sensorimotor feedback loop. For example, the sensorial effect of lateral acceleration is important for curve negotiation (Reymond, Kemeny, Droulez, & Berthoz, 2001). The search for comfort in car manufacturing has sometimes led to a reduction in this kind of feedback and to the well-known phenomenon of risk homeostasis (Wilde, 1982), that is to say an increase in objective risk because of a reduction in subjective risk. The perception mode is fundamental in human–machine cooperation when information is mediatized or enhanced by the machine to create positive interference aimed at improving human performance or the human–machine system performance.

A *mutual control mode* provides the drivers with a judgment of their activities when they approach a limit. This judgment can be either symbolic (e.g., a buzzer) or subsymbolic (e.g., a rumble strip noise). In the first case, the driver must interpret the noise. In the second case, the noise is a familiar sound when the car is approaching the road edge. Several cooperation modes of this type are possible according to increasing levels of intrusion. A *warning mode* just informs the driver of the approach of a limit. An *action suggestion mode* can suggest an appropriate action on a control (e.g., an appropriate vibration of the steering wheel). A *limit mode* can introduce a resistance against a driver's inappropriate action. A *correction mode* can correct the driver's action. All these modes pose the problem of elaboration and maintenance of a COFOR between the situation analysis produced by the driver and that produced by the device with a heavy temporal constraint.

A *function delegation* mode leads to a more continuous intervention of the device. This mode can be *mediatized* when a driver's action turns on a temporary automatic regulation. For example, when the driver brakes heavily and maintains the pressure on the pedal over a certain period of time, the ABS entirely manages the braking, avoiding a skid. With a *control mode* or a *prescription mode,* a high-level driving parameter can be controlled in the medium or long term by the device (e.g., longitudinal control with ACC), the reference being chosen by the driver (control mode) or by a road operator (prescription mode). These types of delegation modes open the way to the well-known automation difficulties of complacency, bypassing, overgeneralization, automation surprise, difficulty of returning to manual control, etc.

Finally, a *fully automatic mode* is envisaged in emergency or very difficult situations and poses the problem of returning to manual driving.

CONCLUSION

In this chapter, I have tried to show that the study of dynamic situations, considered as interactions among human operators, tasks, other operators, and machines, gives us a good opportunity to identify the adaptive properties of human cognition. Some of them are not so salient in laboratory studies where subjects are constrained to exhibit a standard behavior because of the methodological rule to control factors as much as possible. On the contrary, in work situations, people have more degrees of freedom at their disposal.

Facing unexpected situations, but having developed a high level of expertise, human operators are able to (and must) use diverse modalities of cognitive control, sometimes in parallel. However, researchers are dealing with methodological difficulties when trying to identify the control modalities, which can act in parallel and which need intrusive identification means. Confronted with autonomous machines acting in the same environment, human operators need to develop cooperation with them. The model of human–human cooperation may be a good candidate to import useful theoretical constructs into the human–machine cooperation domain. Such an importation has already been done within the context of static situations, with slight temporal constraints. However, the transposition to dynamic situations is not straightforward because it must include cooperation at the subsymbolic levels of cognitive functioning. In-car automation is a good candidate for this kind of study because human–machine cooperation must be integrated into the drivers' routines, without introducing extra load. With the increase in machine intelligence (adaptive power and autonomy), the concept of human–machine cooperation will be of increasing importance within the domain of human–machine (or computer) interaction.

REFERENCES

Amalberti, R. (1996). *La conduite de systèmes à risques* [Controlling risky systems]. Paris: Presses Universitaires de France.

Amalberti, R. (2001). The paradoxes of almost totally safe transportation systems. *Safety Science, 37,* 109–126.

Amalberti, R., & Deblon, F. (1992). Cognitive modelling of fighter aircraft's process control: A step towards an intelligent onboard assistance system. *International Journal of Man–Machine Studies, 36,* 639–671.

Amalberti, R., & Hoc, J. M. (2003). *Cognitive control and adaptation. Lessons drawn from the supervision of complex dynamic situations.* Manuscript submitted for publication.

Andro, M., Chauvin, C., & Le Bouar, G. (2003). Interaction management in collision avoidance at sea. In G. C. van der Veer & J. F. Hoorn (Eds.), *Proceedings of CSAPC '03* (pp. 61–66). Rocquencourt, France: EACE.

Bainbridge, L. (1988). Types of representation. In L. P. Goodstein, H. B. Anderson, & S. E. Olsen (Eds.), *Tasks, errors, and mental models* (pp. 70–91). London: Taylor & Francis.

Card, S. K., Moran, T. P., & Newell, A. (1983). *The psychology of human–computer interaction.* Hillsdale, NJ: Lawrence Erlbaum Associates.

Castelfranchi, C. (1998). Modelling social action for agents. *Artificial Intelligence, 103,* 157–182.

De Keyser, V. (1995). Time in ergonomics research. *Ergonomics, 38,* 1639–1660.

Endsley, M. (1995). Toward a theory of situation awareness in dynamic systems. *Human Factors, 37,* 32–64.

Entin, E. E., & Serfaty, D. (1999). Adaptive team coordination. *Human Factors, 41,* 312–325.
Gibson, J. J. (1986). *The ecological approach to visual perception.* Hillsdale, NJ: Lawrence Erlbaum Associates. (Original work published 1979)
Heath, C., & Luff, P. (1994). Activité distribuée et organisation de l'interaction [Distributed activity and interaction organization]. *Sociologie du Travail, 4,* 523–545.
Hoc, J. M. (1988). *Cognitive psychology of planning* (C. Greenbaum, Trans.). London: Academic Press. (Original work published 1987)
Hoc, J. M. (1989). Strategies in controlling a continuous process with long response latencies: Needs for computer support to diagnosis. *International Journal of Man–Machine Studies, 30,* 47–67.
Hoc, J. M (2001). Towards a cognitive approach to human–machine cooperation in dynamic situations. *International Journal of Human–Computer Studies, 54,* 509–540.
Hoc, J. M., Amalberti, R., & Plee, G. (2000). Vitesse du processus et temps partagé: Planification et concurrence attentionnelle [Process speed and time-sharing: Planning and attentional concurrence]. *L'Année Psychologique, 100,* 629–660.
Hoc, J. M., & Blosseville, J. M. (2003). *Cooperation between drivers and in-car automatic driving assistance.* In G. C. van der Veer & J. F. Hoorn (Eds.), *Proceedings of CSAPC '03* (pp. 17–22). Rocquencourt, France: EACE.
Hoc, J. M., & Carlier, X. (2002). Role of a common frame of reference in cognitive cooperation: Sharing tasks between agents in air traffic control. *Cognition, Work, & Technology, 4,* 37–47.
Hoc, J. M., & Debernard, S. (2002). Respective demands of task and function allocation on human–machine co-operation design: A psychological approach. *Connection Science, 14,* 283–295.
Hoc, J. M., & Lemoine, M. P. (1998). Cognitive evaluation of human–human and human–machine cooperation modes in air traffic control. *International Journal of Aviation Psychology, 8,* 1–32.
Hoc, J. M., & Samurçay, R. (1992). An ergonomic approach to knowledge representation. *Reliability Engineering and System Safety, 36,* 217–230.
Hollnagel, E. (1993). *Human reliability analysis: Context and control.* London: Academic Press.
Klahr, D. (1978). Goal formation, planning, and learning by pre-school problem-solvers or: "My socks are in the drier." In R. Siegler (Ed.), *Children thinking: What develops?* (pp. 181–212). Hillsdale, NJ: Lawrence Erlbaum Associates.
Lazarus, R. S. (1966). *Psychological stress and the coping process.* New York: McGraw-Hill.
Loiselet, A. (2002). *La coopération à distance, discontinue et son soutien: Gestion du référentiel commun interne au travers d'un référentiel commun externe* [Remote and discontinuous cooperation, and its support: Management of the internal common frame of reference by the means of an external frame of reference]. Valenciennes, France: CNRS and Univeristy of Valenciennes, LAMIH.
Loiselet, A., & Hoc, J. M. (1999). Assessment of a method to study cognitive cooperation. In J. M. Hoc, P. Millot, E. Hollnagel, & P. C. Cacciabue (Eds.), *Proceedings of CSAPC '99* (pp. 61–66). Valenciennes, France: Presses Universitaires de Valenciennes.
Løvborg, L., & Brehmer, B. (1991). NEWFIRE—a flexible system for running simulated *fire-fighting experiments* (Report No. RISØ-M-2953). Roskilde, Denmark: RISØ National Laboratory.
McCarthy, J. C., Fallon, E., & Bannon, L. (2000). Dialogues on function allocation [Special Issue]. *International Journal of Human–Computer Studies, 52*(2).
Morineau, T., Hoc, J. M., & Denecker, P. (2003). Cognitive control levels in air traffic radar controller activity. *International Journal of Aviation Psychology, 13,* 107–130.
Pacaux-Lemoine, M. P., & Loiselet, A. (2002). A common work space to support the cooperation in the cockpit of a two-seater fighter aircraft. In M. Blay-Fornarino, A. M. Pinna-Dery,

K. Schmidt, & P. Zaraté (Eds.), *Cooperative systems design: A challenge of mobility age* (pp. 157–172). Amsterdam: IOS Press.

Paterson, E. S., Watts-Perotti, J. C., & Woods, D. D. (1999). Voice loops as coordination aids in space shuttle mission control. *Computer Supported Cooperative Work, 8,* 353–371.

Piaget, J. (1974). *Adaptation vitale et psychologie de l'intelligence* [Vital adaptation and psychology of intelligence]. Paris: Hermann.

Rasmussen, J. (1986). *Information processing and human–machine interaction.* Amsterdam: North-Holland.

Reymond, G., Kemeny, A., Droulez, J., & Berthoz, A. (2001). Role of lateral acceleration in curve driving: Driver model and experiments on a real vehicle and a driving simulator. *Human Factors, 43,* 483–495.

Roth, E. M. (1997). Analysis of decision-making in nuclear power plant emergencies: An investigation of aided decision-making. In C. E. Zsambok & G. Klein (Eds.), *Naturalistic decision-making* (pp. 175–182). Mahwah, NJ: Lawrence Erlbaum Associates.

Simon, H. A. (1983). *Reason in human affairs.* London: Basil Blackwell.

Stanton, N. A., & Young, M. S. (1998). Vehicle automation and driving performance. *Ergonomics, 41,* 1014–1028.

Vanderhaegen, F., Crevits, I., Debernard, S., & Millot, P. (1994). Human–machine cooperation: Toward and activity regulation assistance for different air-traffic control levels. *International Journal of Human–Computer Interaction, 6,* 65–104.

Wilde, G. J. S. (1982). Critical issues in risk homeostasis theory. *Risk Analysis, 2,* 349–358.

Zhang, J., & Norman, D. A. (1994). Representations in distributed cognitive tasks. *Cognitive Science, 18,* 87–122.

8

Transferring Technologies to Developing Countries: A Cognitive and Cultural Approach

Carlos Díaz-Canepa
Pontificia Universidad Católica de Chile

Let me introduce the themes of this chapter with an example.

I was recently invited to devise a test that could assess the cognitive skill of Chilean miners who were to be introduced to a new Finnish semi-automatic system for managing the trains that transport minerals. Accustomed to a three-dimensional system using physical signs, the miners would have to learn to use a computer-based system. A task analysis of a system similar to the one planned for the mine suggested that the three-dimensional system was in several ways more reliable than the computer-based system. First, the schematized work process that was used as a reference for the design of the new tool did not correspond with the actual work process. Second, the new system, which relied heavily on electronic technology, proved to be an unreliable source of information: On many occasions, field operators received contradictory instructions through the computer and radio. Problems were compounded when the company removed external physical signs from the site once the new technology was in place, believing them to be redundant. Finally, the company did not foresee the high level of interaction that, in fact, occurred between the semi-automated systems and the technicians in charge of road maintenance.

A cognitive assessment indicated that the workers hired to run the new equipment had low levels of literacy, as measured by reading comprehension tests, making it difficult for them to work with symbolic support systems. They also scored low on tasks measuring skills that were important for operating the new electronic system, such as analogical and visual–spatial reasoning. Finally, we found that workers were accustomed to thinking in concrete terms, reasoning more with reference to perceptual information or immediate social input than to abstract indicators. The engineers in charge were made aware of the mismatch between workers' abilities and the technical requirements of the new system. Their response was that these difficulties were supposed to be solved by those in charge of the operation of such systems, once they were implemented.

This anecdote exemplifies the kinds of problems that commonly occur when we transfer technologies from the industrialized world to the developing (industrializing) world. The difference in context is complex but unfortunately is scarcely considered. Workers receive inadequate training to deploy the new technology; engineers in charge of the transfer of technology do not communicate with the operators who will be in charge of running it. There have been few attempts to learn from the difficulties that recur in the transfer of technologies. It is consequently a great challenge for those interested in developing countries to define a set of criteria that will aid in the transfer of technologies. To define these criteria within a comprehensive psychological theoretical framework is without doubt a much greater challenge. The present chapter discusses a model for the transfer of technology that specifies the roles that people and technology play in work processes. It involves designing adequate mechanisms of coordination, communication, and supervision, and managing the interaction between the new work systems and their immediate environments during the implantation of new technologies.

The model of technology described in the following pages employs three simultaneous levels of analysis. The first level concerns the formal design of devices, technology, and organizations. The second concerns the rules, procedures, and roles that govern the use of these technologies. A third level refers to the actual deployment of devices, technology, and organizations by users. These three levels of analysis imply a dynamic intertwining of technical, psychological, and contextual factors. I will comment on the relation among these factors from the perspective of the interpretative activity performed by people at work (Díaz-Canepa, 1987). I will also emphasize the heterogeneous and contingent character of the conditions under which technological transfer occurs. The very diversity of situations involved in the transfer of technology triggers the need to establish some minimal criteria for the management of technological transfers. The present chapter will first address the cognitive and cultural questions arising from the introduction of new technologies. I will then briefly revisit the ways the behavioral sciences have answered these questions over the past 30 years, ending with a discussion of the concepts of adaptation and technological appropriation. Finally, I will present a conceptual and pragmatic synthesis that I hope addresses some of the challenges arising in the transfer of new technologies to industrializing countries.

THE IMPLEMENTATION OF IMPORTED TECHNOLOGY

More often than not, the incorporation of technology in developing countries takes place by a process of aggregation (Díaz-Canepa, 2000). By aggregation, I refer to the fact that new technologies, and the procedures attached to them, are simply added to those previously in place. In consequence, there is no continuity in the technological philosophies or the organizational logic employed by the older and newer systems. The difficulties are compounded when the technology is transferred from environments with a different level of technological development and distinct cultural characteristics. Moreover, the recipient countries tend to have heterogeneous sociotechnical formations. Because of the fact that most technologies are transferred to the developing world by aggregation, workers dealing with new technologies usually diverge from formal guidelines to adjust them to their new context of use. Consequently, they operate within systems that are, if not degraded, defined at least by different operational orders and rhythms (Guillevic, 1990). The incorporation of technical changes in the work system—and the ensuing redefinition of rules these changes involve—creates, as a consequence, a high level of friction between the formal procedures and the actual activity of the individuals. Indeed, the structure of values imposed by the new technology puts the structure of values of the organization under a high level of tension (Hebel, 2000). For example, they may collide in their definitions of time, quality, or shared work. This tension is usually handled implicitly and is often conducive to varied practical solutions that, hard to predict, determine the successful adoption of any new technology.

Although the processes of technological transference initiate a regular, and not exclusively technical, flow of communication between suppliers and importers of technology, the cultural variables that determine the functionality of the transferred tools are not commonly discussed. The written documents and diagrams that most donors provide as technical support make two assumptions about their receivers: that people decode information through culture-free or neutral processes, and that they are sufficiently literate to decipher technical writing. Unfortunately, as the Organization for Economic Cooperation and Development (OECD; 2000) has noted, deciphering complex technical manuals may not be possible for a large percentage of the population in many developing countries. In Chile, a country that imports most of the technologies it needs, 85.1% of the population has very poor reading comprehension, and only 1.6% possess reading comprehension levels compatible with a knowledge-based society (OECD, 2000). These results contrast with those attained in industrialized societies: Sweden, 27.8% and 32.4%; U.S.A., 26.6% and 21.1%; United Kingdom, 52.1% and 16.6%; Canada, 42.2% and 22.7%, respectively. Chile's literacy levels more closely approximate those of poorer European countries: Portugal, 78% and 4.4%; Poland, 77.1% and 3.1%; and Hungary, 76.5% and 2.6%, respectively. These antecedents without any doubt put "context-free" technological implantation into question.

One strategy used by large multinational companies to limit the difficulties involved in the transfer of technologies has been to try to re-create the original conditions of use in new contexts. Thus, they form virtual sociotechnical "islands" that are somehow "protected" from their cultural surroundings (Wisner, Pavard, Benchekroun, & Geslin, 1997). Unsurprisingly, these attempts to isolate organizations from their contexts have failed, because workplaces are permeated by the more general characteristics of the communities into which they are inserted, sharing their social structures, climate, and local services. Although some multinationals have persisted in using this strategy, many experiences with technological transfer have confirmed Cole's (1996/2003) observation that there is no such thing as an all-purpose context-free tool.

A HISTORICAL SKETCH

Before advancing a new explanatory model concerning the psychological dimensions of technological transfers, I will briefly review the past 30 years of research on this topic.

During the 1970s and early 1980s, researchers focused on making the psychological study of technological transfers a legitimate field of research. These early researchers, building the theoretical, methodological, and ideological foundations of a new field, found it a challenge to delimit its boundaries. Behavioral scientists interested in this research area entered the field with two research questions. First, they were interested in understanding the factors that could influence the relative compatibility and effectiveness of new technologies. Second, they wanted to evaluate the degree of generality of the models and behavioral rules effective in the developed countries (Chapanis, 1975). From a cross-cultural standpoint, Chapanis stated that, because data were virtually nonexistent, the definition of international standards and the incorporation of modern technology in underdeveloped countries were difficult.

Trying to fill the existing gap, the work directed by Chapanis (1975) covered a wide spectrum of situations. For instance, Daftuar (1975) analyzed the use of indoor spaces in India and the relative adaptability of the Western model of housing, and its related devices, to the Hindu traditional lifestyle. He also studied the level of understanding of English traffic signs shown by Hindu nondrivers and found numerous contradictory (or contrary) interpretations. For example, the sign for *No Horn* was often interpreted as *Blow your horn, please.* In addition, the percentages of correct interpretations of road signs were smaller than those obtained in comparable studies in the West. Identical conclusions were reached in a recent study made in Ghana, West Africa (Smith-Jackson & Essuman-Johnson, 2002), regarding the interpretation of safety symbols presented to industrial and trade workers. This study found wide discrepancies between the users' perceptions of the symbols and their intended meanings.

In the 1970s and 1980s, cognitive psychologists began studying a related problem: how people made assumptions about the operation of a device on the basis of its design. Stereotypes in the assumed "direction of movement" refer to the spontaneous manipulation of a tool (e.g., the shower faucet or the push button of a radio) to achieve a desired effect (e.g., increase the flow of the water in the case of the shower or the volume of music in the case of the radio). The existence of stereotypes is assumed when these spontaneous choices are shared by at least 80% of a determined population. From a psychological point of view, the stereotypical movements are upheld in operational mental representations, originating in part from cultural factors (Sperandio, 1980). These studies (e.g., Díaz-Canepa, 1982; Kroemer, 1975) did not conclusively demonstrate the cultural relativity of such judgments; they did demonstrate that there are significant cultural differences in the intensity of such stereotypes. I will illustrate this finding with an example from my research in Algeria.

The study involved 400 young Algerians, students at Algeria's technical institutes, between 18 and 20 years old, of both sexes, literate and bilingual in Arabic and French. The study presented groups of 100 students with a pencil-and-paper test containing 48 stimuli. Each stimulus presented a particular design of a device that included the window of a control panel (in horizontal or vertical position), the pointer within it (right–left or up–down, according to the position of the window) and a button or a lever that commanded the pointer (placed up or down or left–right, according to the position of the window). Subjects were asked to indicate to which side they have to move the button or the lever in order to move the pointer in a specific direction. The results indicated that, despite sharing direct experience with modern tools, the students did not reach a consensus regarding the operation of simple devices such as those shown in Fig. 8.1. The rectangle represents the window, the black triangle the pointer, and the oval-shaped figure the button. The figure shows the percentage of the sample that indicated the button had to be moved in a specific direction in order to move the pointer in the direction indicated by the arrow.

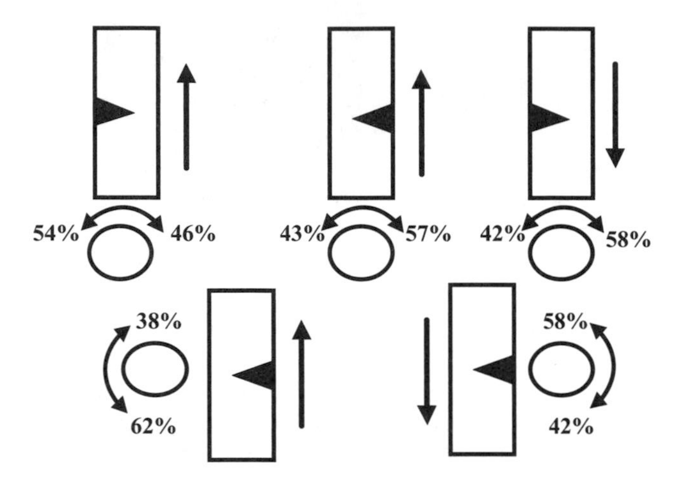

Fig 8.1. Sense of movement applied to the command button, according to movement direction solicited by the control gauge and the configuration of the device, in a population of young Algerians.

A broader study of technology transfers (Lambert, 1979) concluded that their success or failure depended on the economic, geographical, and cultural proximity of operators in the supplier and recipient cultures. These characteristics help determine what Lambert called a culture's "technological mimetic trait"—the tendency of a population to assimilate foreign attributes, objects, and forms of life. Similarly, Wisner et al. (1997) emphasized the importance of social, cultural, anthropometric, geographical, and climatic differences on the success of technological transfers between the so-called first- and third-world contexts.

With an increased awareness of globalization, the focus of research has shifted from the *identification* of cultural discordances in technological transfers to its *management*. For example, Steenhuis and de Brujin (2002) emphasized the learning processes involved in the achievement of a new systemic balance that follows the implantation and use of new technology. They studied learning curves in the transfer of aeronautic technology and showed that the time needed to achieve the productive standards proposed by the suppliers of technology tended to be considerably longer in receiving countries than in supplier countries. They attributed the lag to differences in the conditions in which learning occurred.

The field of action-oriented ergonomics has tackled similar questions. One line of research stresses that worker participation and socialization is crucial to the successful incorporation of new technologies. Kogi (1997), for instance, discussed more than 1,700 cases of successful technological transfers in small- and medium-size companies in Asia and South America.

He identified several ingredients of successful transfer, such as the use of local resources, the consideration of working conditions, and integration of productive objectives. However, he conjectured that participatory and constructive strategies may be limited to medium-size companies, insofar as they possess greater flexibility than larger companies and may have a better knowledge of local resources. Finally, researchers working on problems of technological change and technological transfers have begun to collaborate, identifying the importance of social and organizational variables in the successful management of technologies (Courteney, 2000).

Currently, there are two main approaches in the field. Some researchers favor a normative strategy that emphasizes technical training and the transfer of "simple" technologies, whereas others favor a constructive strategy that evaluates the resources and local dynamics of the receiving country. Those endorsing the first approach talk in terms of the *adaptation* of technology, emphasizing the necessary adjustment of technologies to the user's characteristics (Wisner et al., 1997). Those favoring the second approach speak of the *appropriation* of technology, and point to the active character that users assume while incorporating technology (Guillevic, 1990; Pouloudi, Perry, & Saini, 1999).

In spite of their merits, these more pragmatic approaches to technology have not tackled the theoretical and practical problems posed by culture. On the contrary, they have submerged culture within so-called contextual variables. The identification of cultural characteristics of the population relevant to the importation of new technologies remains a theoretical challenge. We believe that recent developments based on the cognitive theory of Piaget (Gullevic, 1990) and the cultural-historical theory of Vygotsky (Clot, 1999; Cole, 1996/2003; Engeström, 1999; Rabardel, 1995) could help us tackle this issue.

COGNITIVE STATUS OF TECHNOLOGY AND INSTRUMENTAL ACTIVITY

The Meaning of a Device

The subject–device relationship is defined not only by the material qualities of the device—its affordances (Norman, 1991)—but also by the concepts and meanings previously attributed to it, specifically its modes of use. As noted by Cole (1996/2003), the artifacts have a dual conceptual-material nature by virtue of their prior participation in goal-directed human activity. From this perspective, the world of devices is not only a world of physical objects, but also involves a subjective dimension, related to the purposes and ways of using them (Leontiev, 1972/1976). Generally speaking, we can state that:

1. Tools have a multidimensional nature. That is, they appear as technical objects (material and symbolic), they channel meaning, and they mediate our relationship with the environment. They do not exist in isolation, but invoke a context and a structure; they bring to mind the actions associated with their use (implying a representation of their ends), and they possess a range of effective functions that determine the way and the extent to which they affect, have access to, or influence the environment.
2. The context in which individuals' activities take place determines the way a tool is used and the functional value it assumes. The work setting is not only a background, but actually determines how a tool's characteristics—its complexity, affordability, functional rhythm, capacity, and flexibility—affect people's activities. Instruments, for example, are technical objects, but also contain schemes of use developed by the user and/or procedural patterns that have been influenced by other preexisting social schemata (Rabardel, 1995).

Instruments do not, therefore, have a unique and stable meaning but can gradually accrue a range of meanings from a variety of sources: the user's characteristics and previous experiences; the user's prior expectations about the instrument; the consequences of a user's experimentation with an instrument; and the context (social, cultural, and material) in which the instrument is used.

It is through action that a protagonist gives meaning to the object's purpose and modes of use, and it is through action that a device appears as such. At the same time, a device may be used for new purposes exceeding the uses that gave it its identity; it is able to extend or open new action possibilities for the person.

Devices as Mediators

Guillevic (1990) defines an instrument as an "object made by someone who has conceived it and which serves as a mediator between an operator's action and his field of work" (p. 143, translation by the author). In a similar vein, Engeström (1999) observed that the idea of mediation by tools breaks down the Cartesian wall that isolates the individual mind from culture and society. Devices, in this view, have the ability to enlarge an individual's ability to shape reality and influence other human agents. However, at the same time, a device limits or focuses an individual's work by virtue of its specific features and purposes. The concept of a mediator, therefore, has a double meaning: On the one hand it is a facilitator or enabler; and on the other hand, it is a constraint, as someone or something that separates. Although instruments expand an individual's scope of action on the world and on himself or herself (Vygotsky, 1934/1997), they simultaneously create new demands for the development of their activity.

The use of a tool to accomplish work imposes a distance between an actor and the world that restricts the subject's direct experience. It limits the knowledge one can obtain from the object to those indicators provided by an artifact; likewise, an instrument's features condition the object that will be produced through work. Limiting the subject's contact with the object affects the nature, shape, and amplitude of such contacts. In brief, a person using a device is, to a greater or lesser degree, constrained by the actions the artifact encourages or disallows. The features of a tool—its affordances, as it were—are critical factors in the construction and use of cognitive supports that allow people to act on reality. Such supports must to a greater or lesser extent incorporate both generic aspects of reality (independent of the devices used) and aspects that relate to the ends, functions, and expressive modalities of the instruments being used (Luria, 1981).

From Engeström's (1999) perspective, the role of artifacts in mediating the relationship between subject and object of an activity is framed by social dimensions: rules, community norms, the division of labor. From a cognitive point of view, mediation by instruments involves the articulation of at least four levels of representations, schemata, or symbolic structures: those gathered from reality by the operators; those registered in the instruments; those built by operators based on their knowledge, competencies, and objectives regarding the instruments; and those that are inherited or coconstructed through socialization with others.

THE ROLE OF CONTEXT AND CONTINGENCIES

Shared Dynamic Context and Rules

Context plays a particularly important role in the representation of a device's uses. The way instruments are used and the frequency with which they are used depend on their availability in a given situation. The functional value of a device is an attribute that is "culturally defined, historically dynamic, and variable as a function of the situation" (Jordan & Shragger, 1991, cited in Rabardel, 1999, p. 251). However, as Rabardel (1999) points out, it is necessary to establish a difference between the stable functional value of an instrument (its significance) and its situated functional value (its sense), just as the linguistic sign (i.e., a dictionary term) has a meaning that does not necessarily correspond to the sense it assumes in an act of communication.

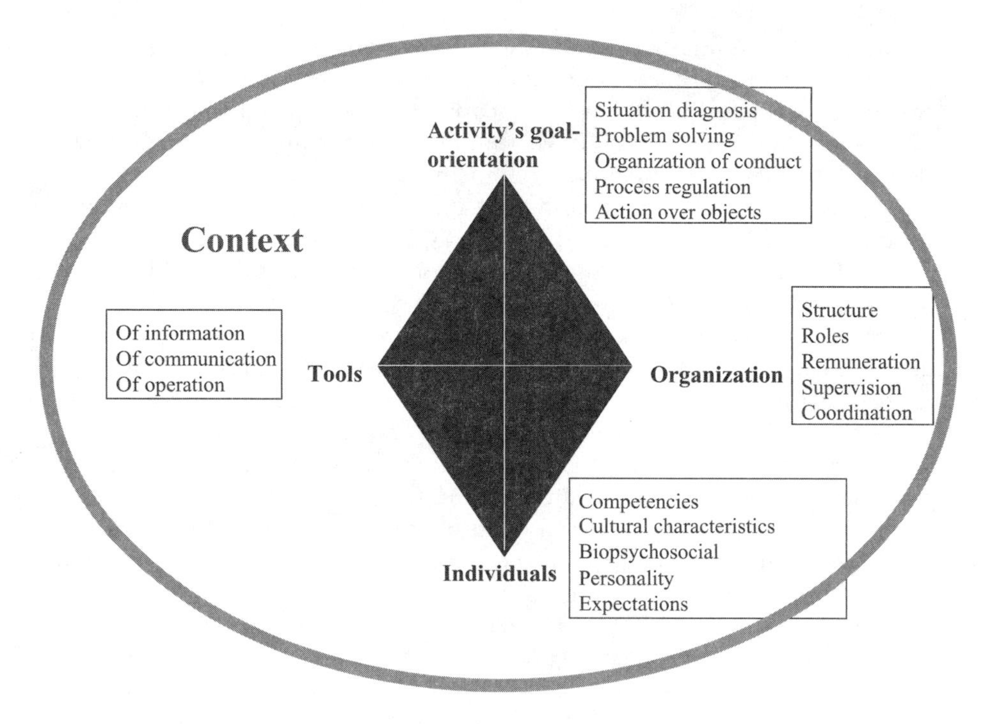

Fig 8.2. Generic model of work systems.

The following section examines the work environment as a context, defined by what we will generically call work systems. As shown in Fig. 8.2, work systems are made up of diverse elements, which in order to achieve their systemic objectives (permanency in time, results, developments, etc.) must generate particular conditions, complementary relations and a global balance.

A work system may be understood in terms of a technology complex (Fleck & Howells, 2001). A technology complex is composed of a series of mutually dependent interactions that sustain the work system: interactions between the tools used in the work field, the principles according to which roles and functions are distinguished within the organization, and the environment into which the organization is inserted. Thus, devices and other elements of the work system (persons, work organization, and rules) help define a dynamic work environment, and come to have, in a manner of speaking, a life cycle of their own (Díaz-Canepa, 2002). These different dynamics gradually give form to particular factual and interpretative contexts. They affect the organization and actualization of cognitive representations relevant to the task, as well as the effective activity of workers.

Transplanting new technologies ruptures a circumstantial balance within the work system. The interactions taking place within the work system, and those taking place between the work system and its environment, can be characterized in phenomenological terms—although such interactions take the form of rules and procedures that incorporate both operative (technical) and social elements. Some of these social regulations characterize the sociotechnical system: roles, behavior regulations, hygiene and security regulations, and criteria for the evaluation of performance. Others refer to a framework of external regulations: the judicial system, relationships with suppliers, clients, neighbors, and broader systems of shared social values and expectations.

Bearing in mind the variety of regulatory elements that bear on the work situation, I will use the term *work rules* to refer to those social and technical principles that regulate people's activity in work situations. These rules can have different levels of formalization, explicitness, and socialization; they can be generated on different occasions and can be expressed by a variety of organizational indicators.

The Purpose and Use of Work Rule

From a Piagetian perspective, systemic rules have an equilibrating and adaptive function. They assume the role of an inter-subjective referent for individual and collective operations. The construction and arrangement of operational rules takes place according to certain underlying logical frameworks, which imbue them with meaning and a realm of application. According to Rabardel (1995), such frameworks can be organized into three major complementary categories:

- *Functional logic*: A framework that describes the functional state of a device.
- *Process logic*: A framework describing the range of actions or transformations that can be achieved by the device.
- *Utilization logic* (instrumental relation): A framework that defines the effectiveness of a device as a means for action.

Distinguishing these logical frameworks allows us to define different prescriptive frameworks, each of which performs a different regulatory function:

1. A framework that describes how the device's components function and how they are to be operated. For example, a steering wheel's functional rule is that it turns in both directions, which are associated to opposite outcomes

(right–left, open–close, etc.). The corresponding operating rule allows it to turn in one direction or another, as desired. The operator, in order to carry out this action, must first ensure that the steering wheel is ready to turn. Then, by turning it in one direction or the other he or she will obtain the expected outcomes (clockwise turn = change in trajectory to the right, for example), making sure that this condition is maintained during the course of the operations carried out with the device.

2. A framework that describes the ways in which an instrument may be manipulated to accomplish a transformation. For example, getting a car to move involves the articulation of turning the key, disengaging the handbrake, pressing the clutch pedal, getting the gearshift to the right position, pressing the gas pedal in coordination with the release of the clutch pedal, guiding the trajectory with the steering wheel, etc.

3. A framework that allows the selection of a path toward a goal, or the choice of a particular instrument to obtain a particular end in a particular way. For example, the same route can be traveled by motorcycle or car, and the decision, in this case, might depend on the volume of the load to be carried, associated cost, available time, etc. In the same way, driving in the city or on the highway, on a rainy or a sunny day, with heavy or light traffic, with or without time pressures, are some situations that undoubtedly require different modes of action in order to achieve the greatest effectiveness in each case.

Rules, Role Representation, and the Regulation of Activity

The attributions that an organization or its workers make about their roles critically affect their ability to carry out an operational representation. According to de La Garza, Maggi, & Weill-Fassina (1998, pp. 415–422) workers encounter two distinct prescriptive fields: one, the "obligation to solve," that is, to obtain a particular solution, where only methods may vary—could be termed a *discretional* prescriptive field. A second, which Maggi and Masino call the "freedom to prescribe our own activity or that of others," describes an *autonomous* prescriptive field. The representation of work roles is determined by the nature of the prescriptive field where the workers participate.

Workers respond to several implicit obligations when performing their jobs (de Terssac & Chabaud, 1990) or interacting with other members of their organization. The redefinition of prescribed rules, through people's activity *in context,* determines a new referential framework, which, in spite of the absence of formality, plays an important regulatory role. The redefined capacity of rules—their regulatory quality—as well as their referential extension, depends extensively on their degree of socialization within the work group as well as on the acceptance they find among supervisors (de Terssac, 1990). Although the processes involving the redefinition of prescribed work rules are usually seen in a positive light, because of their *constructive* and *contextual* nature, it is possible to think of work situations where we can find as many redefined rules as workers performing them.

To illustrate, let's take the example of a study conducted in the subway system in Santiago, Chile. Both the technology and organization of Santiago's subway system, which was established in 1975, originated in France. The operation of subway cars is closely controlled by a set of standards, procedures, and controls that "encloses" the activity of the operators. To use Hatchuel's (1996) terms, the subway conductor's work is organized according to a strongly prescribed system.

During an evaluation of subway conductors' workloads in early 1988, it was observed that the regulations were not being followed. Although conductors were scheduled to drive 330 minutes daily, they completed the trips at an average of 245 and 265 minutes per conductor, though all trains completed their scheduled routes.

An analysis of the conductors' field records and activities provided an explanation for this mystery: Conductors relied on an implicit mechanism of assigning driving times, which included unprescribed breaks after approximately 120 minutes of driving. This alteration had occurred spontaneously and wasn't even registered by the field supervisor in charge of controlling the schedule. A difference in driving times was seen only when "reserve" drivers, whose driving assignments were formally prescribed, took over from scheduled drivers.

This spontaneous regulation was due to the fact that conductors disliked driving for more than 2 hours without a break. This dislike made itself known by an explosive increase in medical leaves of absence—the issue, in fact, that prompted the study.

Although the workers introduced a change to their rhythm of work in order to improve their working conditions, it had a perhaps unintended negative consequence for them—it resulted in an unequal distribution of work. In addition, there was a clear difference in the pattern of medical leaves that different groups of conductors presented, depending on their degree of adaptation to prolonged periods of driving. This differentiation was found to be closely associated with the personal characteristics of the drivers, such as age, length of time in the position, and type of formal education.

This example demonstrates several principles: First, people (in this case, the drivers and their direct supervisors) can and do go beyond the official rules (by, in this case, negotiating new rules in the field). Second, the generation of an informal set of rules signals that the criteria established by the providers of the technology have not been deeply internalized. Third, the

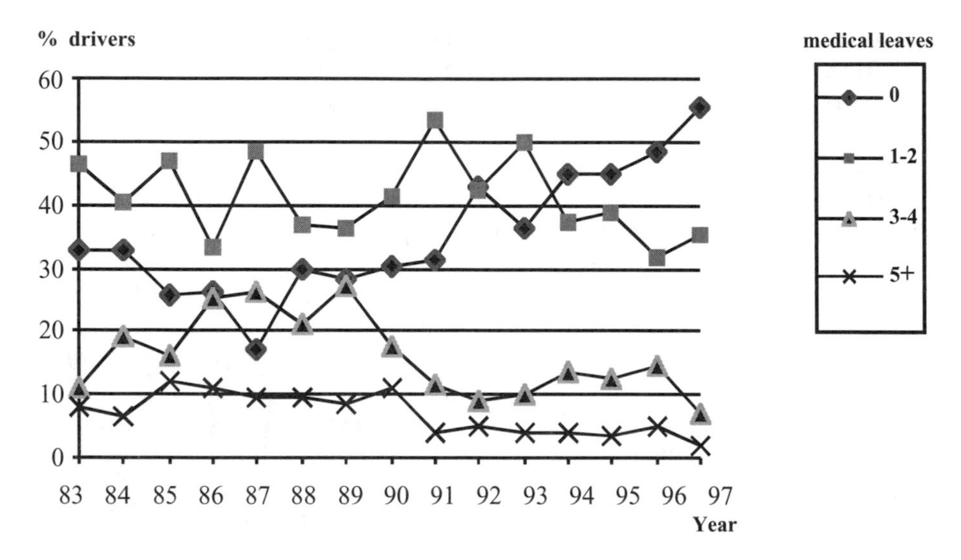

Fig 8.3. Distribution of the percentage of subway drivers according to the number of medical leaves taken by them for each year.

adaptive styles that the people exhibit in work situations may acquire specific forms as they pass through the filter of their personal characteristics.

The conflict between the demands of the driving schedules and the conductors' needs—a conflict that predated our intervention at the end of 1988—was slowly and steadily reduced. As shown in Fig. 8.3, the percentage of drivers who took no medical leave in a given year decreased steadily between 1983 and 1987, whereas, during the same period, the percentage of drivers who took between three and four medical leaves in the year increased. In 1988 this trend began to change: The percentage of drivers who took no medical leave grew significantly, whereas those requesting three or four leaves decreased.

The increasing availability of drivers over time can be explained, in part, by the drivers' greater psycho-physiological adjustment to the new system, but is also due to their active participation in the design and implementation of the driving schedule.

ADAPTATION VERSUS APPROPRIATION

Normally, the extent to which imported technologies are adopted depends on the degree to which new rules and values associated with the technologies are internalized. The success or failure of a technology transfer depends more on the workers' assumption on new norms and principles (so that

TABLE 8.1
Percentage of Subjects Using Substitute Strategies

Substitute Mechanisms	*Written %*	*Unix %*	*Excel %*	*Electronic %*	*Verbal-motor %*
Sales	53.3	13.3	20	—	20
Company Total	67.5	8.1	13.5	5.4	13.5

the machine is used as it was designed) than to the actual effectiveness of the machinery. I have used the term *functional scission* to describe the gap between prescribed and actual uses of imported technologies that is due to a failure of rule internalization (Díaz-Canepa, 2000). Functional scission occurs because, on the one hand, formal systems are implanted; but, on the other hand, informal systems arise, based on the local history of the organization of work or on spontaneous invention. It is common to find situations in which operators carry out different actions to accomplish the identical objective, neither of which conforms to prescribed actions. The strategies tend to be practiced simultaneously, yet are not functionally coordinated and may even interfere with one another.

To illustrate, let's take the example of a study of the work strategies used by operators of a sales management computer system (Díaz-Canepa, Guerrero, Goldfarb, Robles, & Rojas, 1999). The operators were observed while shifting from a Unix-based system to another operating system. Tables 8.1 and 8.2 show the distribution of two strategies used during the transition: strategies to compensate for or to substitute for the use of the new technology. The data were collected using interviews with operators and intermediate-level workers. In each interview, workers reported when they used compensatory and substitute strategies. The tables show the distinct forms of the strategies acquired, considering that many users employed more than one strategy to compensate or substitute the use of the new technology.

The data show a tendency for workers to resort increasingly to primary working modalities (paper and pencil, verbal–motor), which contrasts with the potential of the available technology. From a Piagetian perspective, as workers encountered difficulties in using the new computational tool, they experienced a loss of structure and reverted to more archaic schemata. We

TABLE 8.2
Percentage of Subjects Using Compensatory Strategies

Compensatory Mechanisms	*Written %*	*Unix %*	*Excel %*	*Electronic %*	*Verbal-motor %*
Sales	46.6	13.3	6.6	46.6	53.3
Company Total	21.6	24.3	5.4	32.4	43.2

found that certain compensatory schemes, such as resorting to alternative electronic support (calculators specifically), were the function of two kinds of mismatches. First, some needed calculation procedures were not included in the new system. Second, the aims of the formal work system were different from those of the salespeople who used them. Whereas the former were focused on the volume of the operations to be carried out, the latter was structured based on the commercial margins of operations.

Successful appropriation, then, occurs when imported elements can be integrated to some degree with a preexisting work situation and/or to the users' previous schemata. The appropriation process can accordingly be understood as one modality in which people "attempt to integrate the new tool's utilization within the set of schemata previously constructed;" this process "would allow the operator, through successive regulation, to dominate the tool in a specific context" (Guillevic, 1990, p. 147, author's translation).

According to Guillevic, appropriation of a tool results from the accommodation of the operators' cognitive schemata to the new instruments' functional characteristics. We think differently. From our viewpoint, to the extent that appropriation is used to refer to a process guided by the features of the new context and by the operators' actual activity, appropriation also involves—to a critical extent—the assimilation of the tool to a new activity or purpose.

Appropriation can take a variety of forms, but not all of them are necessarily favorable to the successful transfer of technology. We can distinguish between two broad modalities of appropriation. One modality is *appropriation by means of fusion*, which operates by substituting the old devices for new ones, but maintaining the ways of operation associated with the old devices (for example, using computers as typing machines). This modality of technological incorporation in a work context often results in the adoption, by analogy, of certain formal and ritual aspects of what is prescribed. This form of appropriation could be explained, from a cognitive viewpoint, as an overruling of assimilation over accommodation.

A second modality is *appropriation by means of active restructuring*, which operates through the reformulation of the implanted systems' objectives and rules of utilization, according to the characteristics, needs, and behavior of the new context. In order to be maintained and become a referential framework shared by the work group, this second modality of instrumental appropriation might require more conscious levels of action by workers than would other modes of internalization. It would imply, to a great extent, the socialization of participatory dynamics to construct the processes and rules guiding work. Despite the constructive nature of this mode of appropriation, it is also true that its viability requires that new work rules be sufficiently internalized and that the transformational processes wherein people and tools interact be comprehended.

CONSIDERATIONS IN MANAGING TRANSFERS OF TECHNOLOGY

To present a different approach to the predominant way of dealing with technological transfers I distinguished two different modalities of technological and organizational incorporation. These modalities have been schematically categorized as adaptation and appropriation. Whereas adaptation is a process driven by the provider of the new technology, appropriation is based on the ends, mental representations, competencies, and contexts that exist in the receiver's work situation. The locus of control in adaptation is external and hierarchical. The locus of control in appropriation is internal and socialized. Adaptation emphasizes the role of memorization in the transfer of knowledge; appropriation calls for the reelaboration of situated learning.

Although there are undoubtedly advantages to the process of appropriation, it requires an adequate system of support. In order to facilitate appropriation, one needs to identify and monitor indicators of the process of technological incorporation: error rates, compensatory or substitute activities developed by workers, etc. In addition, it helps to be able to characterize the relationship between individuals and devices in terms of functional logic, process logic, or utilization logic (as defined by Rabardel, 1999). Knowing how tasks are carried out (prevailing work logic) and how supervision is accomplished (the hierarchical operative relations) makes it easier to identify what and how workers need to learn, and to relate the task of learning with the accomplishment of a purpose. To plan an appropriate learning curve, as Steenhuis and de Brujin (2002) suggest, means to articulate a method of incorporating an instrument and to take on the more or less explicit demand for socialization implied in this process (Kogi, 1997). Managing a transfer of technology also requires setting goals over time, so that workers can progressively incorporate a new technology into existing work rhythms. This presupposes adequate resources to formulate such goals and prescriptions and to supervise and assist the process of change. Finally, it is important to identify the impact a new technology may have on the distribution of roles, on decision-making power (Rasmussen, Brehmer, & Leplat, 1991), and on cognition (Hutchins, 1995; Salomon, 1993).

CONCLUSION

The study of technological transferences has had an erratic history because, in large part, of the wide variety of situations that can be termed *transfers of technology*. It has been difficult to construct models general enough to accommodate this diversity of situations, yet concrete enough to facilitate problem solving. The cognitive-instrumental approach has become a powerful model because it draws from historical–cultural studies of cognition and work. It has facilitated a new understanding of the relationship between work and cognition, in which work is recognized as a complex constructive phenomenon that cannot be considered apart from its context and from its mediating role.

From within this framework, four key components to the incorporation of technology through transfer can be identified:

- The purposes that give place and sense to the activity.
- The context of reception.
- The characteristics of the devices in question.
- The modalities of technological incorporation.

We maintain the boundaries underlying technical and organizational systems, which are defined as the technological and organizational margins of flexibility resulting from the nature of such instruments as things made according to concepts. We also uphold the constructive nature of learning and the sociability inherent in sociotechnical systems. As Zucchermaglio (1995, p. 67) states, "this constructive activity is not completely free and creative, but depends strictly on the specific constraints and characteristics of the environment." However, the instruments and systems transferred do not necessarily possess functional characteristics that guarantee by themselves the results obtained in their contexts of origin, because those same contexts are, in part, responsible for the functions of these tools, particularly through the activity of workers in those contexts. Such conditions critically influence sociocognitive processes of meaning-making and action and are set by the people called on to use these new technologies, constructing what could be called contextual-technology. In a sense, we seem to be far removed from the hope of a cosmopolitan technology expressed 3 decades ago (Chapanis, 1975).

However, it is imaginable that traditional strategies of training and selection—frequently associated with adaptation—can help establish the rudiments of technological transfer, but only when they are articulated in terms of the purposes and context of the activity. Accommodation is a necessary phase that generates new representations with which to describe, explain, and communicate experiences (Piaget, 1967/1997, p. 334). They themselves promote the construction of new assimilation schemata. Such accommodation must not remain "on the surface of things" (Piaget, 1967/1997, p. 335), but must be allowed to dig "under the chaos of appearances [searching for] regularities to allow for effective experimentations" (Piaget, 1967/1997, p. 338). The construction of elaborated and shared representations of the situation may hasten the priority of context-based purposeful activity over the cold logic of technologies. When peoples' purposeful action becomes the focus of analysis, it not only facilitates successful technological transfers, but also accords a significant role to local innovation and design activity in the work performed in industrially developing countries.

REFERENCES

Clot, Y. (1999). *Avec Vygotski* [With Vigotski]. Paris: La Dispute.

Chapanis, A. (1975). *Ethnic variables in human factors engineering.* Baltimore: Johns. Hopkins University Press.

Cole, M. (2003). *Cultural psychology: A once and future discipline.* Cambridge: Harvard University Press. (Original work published 1996)

Courteney, H. (2000). The new frontier issues for cognition technology. *Cognition, Technology & Work, 2,* 142–153.

Daftuar, N. (1975). The role of human factors engineering in under-developed countries, with special reference to India. In A. Chapanis (Ed.), *Ethnic variables in human factors engineering* (pp. 91–114). Baltimore: Johns Hopkins University Press.

de La Garza, C., Maggi, B., & Weill-Fassina, A. (1998). Temps, autonomic et discrétion dans les tâches de maintenance d'infrastructure ferroviaire. [Time autonomy and discretion in the maintenance of railway infrastructures] Actes du XXXIII ème Congrès de la SELF, pp. 415–422, Paris.

de Terssac, G. (1990). La polyvalence redéfinie par les interessés [Versatility redefined by those who are involved]. In M. Dadoy, C. Henry, B. Hillau, G. de Terssac, J. F. Troussier, & A. Weil-Fassina (Eds.), *Les analyses du travail, enjeux et formes* [Work analysis, dilemmas and method]. Paris: Cereq, coll. des études.

de Terssac, G., & Chabaud, C. (1990). Référentiel opératif commun et fiabilité [The human factors of the reliability in complex systems]. In J. Leplat & G. de Terssac (Eds.), *Les facteurs humains de la fiabilité dans les systèmes complexes* [Referential of common operation and reliability] [English translation], [pp. 111–139]. Toulouse, France: Octares.

Díaz-Canepa, C. (1982). *Etude expérimental des estéreotypes du sens du mouvement pour des dispositifs de contrôle et de commande* [Experimental study of direction of movement stereotypes for control and direction artifacts]. Alger, Algeria: Institut National d'Hygiène et Securité. Ministere du Travail d'Algerie.

Díaz-Canepa, C. (1987). Rol del factor cultura y de las representaciones mentales en las relaciones con los instrumentos técnicos [Culture factor and mental representations role in relationship with technical tools]. *Revista Chilena de Psicología, IX*(1), 9–14.

Díaz-Canepa, C. (2000). Facteurs personnels, charge de travail et fiabilité dans le poste de conduite du métro de Santiago du Chili [Personnal variables, work load, and reliability in Santiago of Chile's subway conduction]. In *Ergonomics and human factors in train transport: Proceedings from SELF's satellite meeting,* Toulouse, France.

Díaz-Canepa, C. (2002). Enfoque sistémico en análisis del trabajo: Algunos elementos teóricos y una ilustración empírica [Systemic approach in work analysis: Some theoretical elements and an empirical illustration]. *Psykhé, 11*(2), 43–53.

Díaz-Canepa, C. (Dir.), Guerrero, M. F., Goldfarb, D., Robles, J., & Rojas, J. (1999). *Problemas y estrategias adaptativas en un proceso de implantación informática en la empresa Gallyas S.A.* [Problems and adaptative strategies on an information system implantation in Gallyas S.A. company]. Santiago, Chile: EPUC.

Engeström, Y. (1999). Activity theory and individual and social transformation. In Y. Engeström, R. Meiettinen, & R.-L. Punamäki, *Perspectives on activity theory* (pp. 19–38). Cambridge, UK: Cambridge University Press.

Fleck, J., & Howells, J. (2001). Technology, the technology complex and the paradox of technological determinism. *Technology Analysis & Strategic Management, 13*(4), 523–531.

Guillevic, C. (1990). L'appropiation cognitive de l'outil [Cognitive approriation of tool]. In J. Leplat & G. de Terssac (Eds.), *Les facteurs humains de la fiabilité dans les systèmes complexes* [Human factors in complex system's reliability] (pp. 142–157). Toulouse, France: Octares.

Hatchuel, A. (1996). Cooperation et conception collective: variété et crises des rapports de prescription. [Collective cooperation and design: Variability and crises in rule setting relations]. In G. de Terssac & E. Friedberg (Eds.), *Cooperation et conception* [Cooperation and design] (pp. 101–121), Toulouse, France: Octares.

Hebel, M. (2000). Human values and the management of technological change. *Cognition, Technology & Work, 2,* 106–115.

Hutchins, E. (1995). *Cognition in the wild.* Boston: The MIT Press.

Jordan, D., & Shrager J. (1991). The role of physical properties in understanding the functionality of objects. In K. Hammond & D. Gentner (Eds.), *Proceedings from the Thirteenth Annual Conference of the Cognitive Science Society* (pp. 179–184). Hillsdale, NJ: LEA.

Kogi, K. (1997). Ergonomics and technology transfer into small and medium sized enterprises. *Ergonomics, 40*(10), 1118–1129.

Kroemer, K. (1975). Muscle strength as a criterion in control design for diverse populations. In A. Chapanis (Ed.), *Ethnic variables in human factors engineering* (pp. 67–90). Baltimore: Johns Hopkins University Press.

Lambert, D. C. (1979). *Le mimetisme technologique du tiers-monde* [Tecnological mimecry of third world]. Paris: Ed. Economica.

Leplat, J. (1997). *Regards sur l'activité en situation de travail. Contribution à la psychologie ergonomique* [Perspectives on activity in work situation. A contribution to ergonomic psychology]. Paris: Le travail Humain, PUF.

Leontiev, A. (1976). *Le developpement du psychisme* [Development of the mind]. Paris: Ed. Sociales. (Original work published 1972)

Luria, A. R. (1981). *Language and cognition.* Washington, DC: Winston.

Norman, D. (1991). *Cognitive artifacts in designing interaction. Psychology of human computer interface.* Cambridge, UK: Cambridge University Press.

OECD (2000). *Literacy in the information age: Final report of the international adult literacy survey.* Canada: OECD Statistics.

Piaget, J. (1977). *La construction du réel chez l'enfant* [The construction of reality in the child]. Neuchâtel: Delachaux and Niestle. (Original work published 1937)

Pouloudi, A., Perry, M., & Saini, R. (1999). Organizational appropiation of technology: A case of study. *Cognition, Technology & Work, 1,* 169–178.

Rabardel, P. (1995). *Les hommes et les technologies* [Man and technology]. Paris: A. Colin.

Rabardel, P. (1999). Le langage comme instrument? Élements pour une théorie instrumentale élargie [Language as a tool?: Elements for an extended instrumental theory]. In Y. Clot (Ed.), *Avec Vygotski* [With Vigotski], Paris: La Dispute.

Rasmussen, J., Brehmer, B., & Leplat, J. (Eds.). (1991). *Distributed decision making: Cognitive models for cooperative work.* London: Wiley.

Salomon, G. (Ed.). (1993). *Distributed cognitions: Psychological and educational considerations.* Cambridge, UK: Cambridge University Press.

Smith-Jackson, T. L., & Essuman-Johnson, A. (2002). Cultural ergonomics in Ghana, West Africa: A descriptive survey of industry and trade workers' interpretation of safety symbols. *International Journal of Ocupational Safety and Ergonomics, 8*(1), 37–50.

Steenhuis, H. J., & de Brujin, E. (2002). Technology transfer learning. *Technology Analysis & Strategic Management, 14*(1), 57–66.

Sperandio, J.-C. (1980). *La psychologie en ergonomie* [Psychology in ergonomics]. Paris: PUF.

Vygotsky, L. S. (1997). *Pensée et langage* [Thought and language]. Paris: La Dispute. (Original work published 1934)

Wisner, A., Pavard, B., Benchekroun, H., & Geslin, P. (1997). *Anthropotechnologie. Vers un monde industriel pluricentrique* [Anthropotechnology: For a multipolar world]. Toulouse, France: Octarès.

Zucchermaglio, C. (1995). Organizational and cognitive design of learning environment. In C. Zucchermaglio, S. Bagnara, & S. Stucky (Eds.), *Organizational learning and technological change*, NATO ASI Series. New York: Springer.

IV

INTELLIGENT TECHNOLOGIES AND TECHNOLOGICAL INTELLIGENCES

9

Technologies for Working Intelligence

David D. Preiss
PACE Center at Yale University
Pontificia Universidad Católica de Chile

Robert J. Sternberg
PACE Center at Yale University

Some years ago, in one of the authors' graduate school courses, a student was struggling to explain the nature of a study he wished to implement to a professor. At some point in the discussion, he was asked to put in plain words how the variables involved in the experiment would materialize as rows and columns in a spreadsheet. The professor explained that "spreadsheet representations" illustrated the nature of an experimental study better than a number of verbal statements could. Thus, instead of following up their discussion about the theoretical hypotheses behind the study, they began to discuss the study in terms of how its variables could be represented in a computer spreadsheet. Eventually their communication became more fluent, and they were able to clarify the nature of the study.

This example is illustrative of one of the multiple roles played by technologies. Indeed, the ways a computer spreadsheet supports data analysis are manifold. First, as the professor in the graduate class observed, a computer spreadsheet is an adequate medium of representation. At least for this professor, a computer spreadsheet provided a better model of a proposed study than did the sketches the student was eagerly drawing on a sheet of paper. Yet, there are additional ways that a computer spreadsheet provides intellectual help. In a spreadsheet format, the information becomes easier to manipulate. We can produce and combine new variables in a matter of seconds. Moreover, it makes the information susceptible to manipulation by a computer. Thus, we can entrust to the computer the lengthy analyses we would have needed to do on our own. This way, we can focus our cognitive resources on the more substantive or interpretive aspects of data analysis.

In this chapter, we would like to show that many of our everyday tasks intellectually benefit from some sort of technological help, just as scientific reasoning benefits from the use of computer spreadsheets. In our culture, those aids may be a paper, a calculating machine, or a computer, and, in particular, the signs we manipulate by using each one of these artifacts. In other cultures such aids may be of a quite different nature. For instance, in the pre-Columbian Inca culture of the South American Andes, one of the prevalent technologies was a complex set of ties called *quipu,* whose role was to facilitate counting tasks and imperial accounting. The name quipu comes from the *Quechua* language and means knot (Urton, 2003). The quipus consisted of a long rope from which sets of cords hung. Knots were made in the cords to represent numerical values. It is worth mentioning that, although the Inca Empire did not elaborate a written language, the quipus were maintained as historical records. Indeed, it has been suggested that the quipus not only were used to record statistical information but also to record narrative accounts. Students of the quipu have been able to decipher the methods the Incas used to record statistical data, although they have been unable to determine what objects were, in fact, being counted by a particular arrangement of cords and how the quipus represented narratives or myths. In fact, the quipu are not considered an accomplished writing system (Urton, 2003).

As the Incan quipus and computer spreadsheet illustrate, not only simple cognitive tasks but also very complex ones, such as mathematical computations or scientific reasoning, benefit from the use of technologies. The main goal of this chapter is to discuss, first, how technologies relate to human intelligence and, second, how a consideration of technologies influences the conceptions we hold of human intelligence. To advance this discussion, the chapter contains two main sections. In the first section, "Conceptions on the Nature of Technology," we distinguish a *psychologically informed view* of technology from what could be called a *conventional view* of technology. The conventional view considers technology primarily from the perspective of the changes it produces in the physical environment. The psychologically informed view is sensitive to the intellectual consequences of the use of technology. Because it considers the intellectual aspects of technology, the psychologically informed view encompasses both material and symbolic technologies, thereby enlarging the array of things we deem technological. Indeed, this viewpoint also encompasses symbolic inventions such as mathematics and written language under the heading of technology. In the second section, "A Revised Definition of Intelligence," we demonstrate that theories that view intelligence as static—for instance, the theory of intelligence as a general factor—do not take into consideration the interactive processes involved in the use of technologies. In short, we claim that a psychological consideration of technology can expand our view of intelligence as dynamic, culturally shaped, multiple, and distributed. In closing the chapter, we elaborate on these attributes. In other words, this chapter argues that a clarification of the psychological side of technology entails a review of the artifact-dependent nature of intelligence.

Let us review, then, what has been written about the psychological side of technology in order to address later what we mean by the artifact-dependent nature of intelligence.

CONCEPTIONS ON THE NATURE OF TECHNOLOGY

Let us turn to the definition given by the *Encyclopedia Britannica*, which aims to capture changes in the prevailing definition of words. It currently provides what we have called a conventional view of technology: "the application of scientific knowledge to the practical aims of human life or, as it is sometimes phrased, to the change and manipulation of the human environment" (Encyclopædia Britannica Online, 2004a). It also defines a "tool" as

> An instrument for making material changes on other objects, as by cutting, shearing, striking, rubbing, grinding, squeezing, measuring, or other process. A hand tool is a small manual instrument traditionally operated by the muscular strength of the user; a machine tool is a power-driven mechanism used to cut, shape, or form materials such as wood and metal. Tools are the primary means by which human beings control and manipulate their physical environment. (Encyclopædia Britannica Online, 2004b)

These definitions emphasize the following characteristics. First, technology serves human beings in their attempts to adapt to the environment. Second, technologies are material objects that produce material changes on other objects or on the physical environment. Third, technologies involve the application of knowledge to practical ends. According to this point of view, people see technology as either the application of knowledge or as some sort of physical aid instrumental to pursuing a practical goal.

Here, we intend to challenge this conventional view by advancing a psychologically informed view of technology. This task involves two steps. First, we discuss the intellectual nature of artifacts (or tools) as such. Second, we comment on different kinds of human–artifact interaction. In doing so, we join the arguments made by others. For instance, Nickerson (this volume) too advances a view of technology that differs from the conventional one. He claims that tools may be classified according to the nature of the abilities they amplify: motor, sensory, or cognitive. Cognitive tools "include symbol systems for representing entities, quantities, and relationships, as well as devices and procedures for measuring, computing, inferencing and remembering" (Nickerson, this volume). Although he acknowledges that the distinctions between motor, sensory, and cognitive tools are not sharp, he does not reduce technology to muscle or power-driven tools. Concerning the origin of cognitive tools, he proposes that "much cognitive technology has developed in more or less the same way that the ability to stand, to walk, to run develops in the child: People have done what comes naturally as they have tried to extend their abilities to cope with the problems and to respond to the challenges that life presents" (Nickerson, this volume).

In considering the psychological side of technology, Cole and Derry (this volume) propose that artifacts are not exclusively made of material components. According to Cole and Derry, each artifact includes both a physical and a psychological dimension. Artifacts, the founding blocks of technology, are simultaneously ideal and material and can be classified by the nature of this mixture: Some artifacts are more material than others (such as axes and bowls); others are more ideal than material (such as narratives). Artifacts are material as they are part of the physical world. They are ideal (or psychological) as they are also part of goal-oriented human activities. Indeed, artifacts are initially designed with a purpose in mind. That is, they are invented so we can do something with them. For instance, a computer keyboard is a practical device for writing and computer programming. Second, artifacts are given actual use in the context of goal-oriented actions. For instance, we use a computer keyboard to write a letter, a paper, or a novel, or alternatively to perform calculations in a computer spreadsheet. We do not use a computer keyboard to play a symphony for the piano. In fact, one of the most important consequences of the use of artifacts in goal-oriented activities is that, although artifacts help the user to perform adaptive tasks, they also force the user to adapt to the way artifacts were designed (Díaz-Canepa, this volume). Thus, when we use a keyboard, we have to overlook the alphabet and adjust our typing to the canonical QWERTY ordering of the keyboard (which actually predates computers and was previously used on typing machines of all sorts since the Sholes & Glidden Type Writer was invented and sold from 1874–78). This way, artifact use always involves harmonizing the user's needs with the artifact's design.

Therefore, acting in an environment saturated by artifacts has important cognitive implications. Because artifacts impinge a design on the user, artifact use not only involves application of knowledge, it also involves the transmission of knowledge. Each artifact embodies a specific knowledge set that is actualized by whomever makes use of it. To illustrate, let us think of a car. When driving a car one has to bear in mind the rules governing its operation. Additionally, one has to consider the meaning of the signals monitoring its performance. One cannot change the way things work or disregard the signals' warnings. Moreover, the car is one's immediate environment. Part of this environment is, obviously, physical. However, another part is symbolic. Thus, one relates to the more extended environment—the street, other cars—through the physical and symbolic mediation of the car's features. If one crashes, the weight of the car, vis-à-vis the weight of the object with which one collides, will moderate the magnitude of the impact. Of course, some other aspects of design, such as the availability of air bags, will have a consequential effect in protecting or harming the driver. An equally relevant factor is, of course, the speed of both the car and that of the object it hits. Still, one knows the speed of a car only through its speedometer. Thus, to appropriately monitor the car's speed, an understanding of Arabic numerals is needed, as they are commonly used in speedometers. Additionally, one must be able to match a reported speed to an imagined speed and, especially, to the time it will take to reduce the car's speed to a value of 0.

In short, the rules and signs of an artifact embody a particular form of knowledge they impose on the user. As implied in the car example, both signs and rules are intertwined and the successful operation of any machine involves following the rules of the machine and decoding its symbols' warnings appropriately.

Technological Kinds

Up to this point, we have highlighted the general features of those objects deemed technological. Next, we comment on the impact of these artifacts on human abilities. To begin, we draw on the distinctions made by Salomon and Perkins (this volume) regarding technological influence. They distinguish three forms of technological influence: effects *with*, effects *of*, and effects *through* technology. Effects *with* technology refer to the direct impact a technology has on a particular skill, such as the impact of working *with* a word processor on writing. Effects *of* technology refer to the positive and negative effects that remain after the technology has been removed, such as the cumulative effect *of* using a word processor on writing when we write without one. Effects *through* technology refer to the reshaping and reorganization of a general cognitive activity by technology, such as the influence of Arabic numbers on mathematical reasoning, which we will discuss later. To these kinds of technological influence we add the impact of technologies *throughout* everyday life, which refers to the fact that some technologies—infrequently used but quite prevalent within a population—pervade a significant number of everyday practices.

Salomon and Perkins's distinctions explain how alternative kinds of technological influence may work. We would like to complement their insightful taxonomy of technology with a model that takes as its focus the way

TABLE 9.1
Kinds of Technological Influence

Technologies	*Highly Disseminated*	*Scarcely Disseminated*
Frequently Used	Radical technologies "Impact through"	Expert technologies "Impact with"
Infrequently Used	Pervasive technologies "Impact throughout"	Experimental technologies "Impact of"

technology acts in everyday life; see Table 9.1. Two dimensions are key to differentiating between types of technological influence (Preiss & Sternberg, 2003). These are, first, *frequency of use* and, second, *technological dissemination.* Frequency of use refers to how recurrent the use of a particular technology is in everyday life. Technological dissemination refers to how widespread the use of a particular technology is. These two matter as they refer to the temporal and spatial incidence of technologies. We propose that the effect of technology on human abilities is relative to its effective use and dissemination. Technological impact can be classified according to these two dimensions.

Thus, we distinguish among:

- *Experimental technologies*—less frequently used and scarcely disseminated technologies, as in specific experimental technologies applied to reduced segments of the population. Most of their impact relates to experimental situations that test the effect of a particular technology on a particular skill. This kind corresponds to what Salomon and Perkins (this volume) call the impact *of* a technology.
- *Expert technologies*—frequently used and relatively less disseminated technologies, such as the technologies specific to a particular craft. Most of the impact of expert technologies relates to settings where expert operators perform a complex task with the aid of particular technologies—for example, the technologies pilots use in the cockpit that allow them to distribute cognitive load and to perform complex computations across a technologically rich environment (Hutchins & Klausen, 1996). This kind relates more or less to what Salomon and Perkins (this volume) call the impact of working *with* a technology.
- *Pervasive technologies*—less frequently used and highly disseminated technologies, such as intelligence tests or achievement tests. These technologies have an impact on all facets of everyday life, as they permeate a number of contexts, from school to work. Their impact is mediated by these practices and is relative to the societal value awarded to the technology in question. The greater their value, the more societal practices they will permeate. Let us consider a paradigmatic example: achievement tests. In

a number of places (Grigorenko, Jarvin, Niu, & Preiss, in press; Sternberg, 1996), these tests control access to better educational opportunities. Accordingly, people spend long hours developing the skills valued by these tests, fostering these skills before others. However, these skills are not fostered as a result of the tests, but as a consequence of the social practices they trigger.

• *Radical technologies*—frequently used and highly disseminated technologies, such as the alphabet and the Arabic numeric system. These technologies are so frequently used that they become a part of our mindset: For example, when people learn a particular written script, they not only start to approach language from a distinctive *literate* perspective (Olson, 1994, 1996) but also their basic cognitive skills, such as memory, are affected by the nature of the script they manage (Cole, 1996; Scribner & Cole, 1981). This kind of impact relates to what Salomon and Perkins (this volume) call impact *through* technology.

We discuss the nature of these technologies in more detail later.

Experimental Technologies

Let us start by commenting on experimental technologies, as they are the most commonly treated in the psychological and educational literature. These days, the technologies usually discussed in this literature refer to computer or Internet-related tools, yet under the heading of *experimental technologies* we refer to all artifacts that receive an experimental treatment. Experimental technologies are applied to a number of people and, commonly, for a limited time. Although we have information about the impact of these technologies in controlled situations, we are less informed of the nature of their impact after they are transferred to everyday life. Therefore, some technologies that work as expected in experimental trials may have different consequences in real life where they have to be adapted to different individuals and groups. For instance, it has recently been argued that some sophisticated and knowledge-rich computer tools do not have uniform effects when implemented in classrooms that comprise students of different ethnicities. These tools are not sensitive to the cultural backgrounds of their users, in particular, to the cultural backgrounds of those belonging to minority populations that may not benefit equally from them (Lee, 2003).

Lajoie (this volume) makes the case that computer tools are not designed based on a deficit model but rather on an amplification approach. In general, computer tools are designed to increase existing cognitive skills, not to replace a missing function. However, new tools may soon qualitatively expand our cognitive reach as well. For example, it was recently reported that after researchers implanted electrodes in the parietal reach region of the brains of monkeys, a computer was able to decode the patterns of neuronal activity recorded by the electrodes and to determine the direction in which a monkey was planning to, but did not actually, reach (because they were not rewarded if they reached). The authors reported that after 2 months of practice, the computer accurately predicted the intended direction of the monkey's reach as much as 67%, versus 12.5% by chance (Musallam, Corneil, Greger, Scherberger, & Andersen, 2004). As researchers begin possibly to translate the neuronal basis of human intentions to particular computer applications, these advances forecast an optimistic future for the development of very complex neural prosthetics.

Concerning their impact on more traditional intellectual skills, Salomon and collaborators have argued that some computer tools help challenge a learner to progress within his or her zone of proximal development (Salomon, Globerson, & Guterman, 1989). The zone of proximal development is the difference between what a learner can do alone and what she or he can do with appropriate guidance (Vygotsky, 1978). Although Vygotsky assumed that a human partner would provide this guidance, Salomon and collaborators propose that computers can also foster competencies by facilitating the internalization of external guidance. Their criterion to define a successful internalization is improved competency in the absence of external guidance. In order to test the hypothesis that a computer may act as a mentor, Salomon and his associates designed a software program called Reading Partner, which targeted reading skills (Salomon et al., 1989). Reading Partner provided basic reading principles to aid an adolescent reader to comprehend texts better, accompanied by online metacognitive guidance during the reading process. In short, they showed that seventh graders using the software reported more mental effort in reading, showed better metacognitive skills, improved their reading comprehension, and transferred their metacognitive skills to an adjacent skill: writing (Salomon et al., 1989). Reading comprehension is usually considered a good index of intelligence, in particular, of crystallized intelligence (Mackintosh, 1998). As reported in this study, not only was reading comprehension improved by help embedded in a computer program, the students also transferred the acquired metacognitive skills underlying this improvement to a different domain, in this case, writing. As noted by the authors, the skills provided by means of computer tutoring were, therefore, internalized: The students read and wrote better after the software was removed.

Salomon and collaborators' Reading Partner is one of many technologies with the capacity to target and improve specific skills. Indeed, as noted, the educational literature abounds with examples of computer-based educational technologies (Lajoie, 2000; Lajoie & Derry, 1993; Reinking, 1998). Although the technologies themselves are not easily transferable to everyday situations, some generalizations can be made about their implementation. Thus, approaching the issue of the impact of multimedia technologies, Mayer (2003) detected four effects that are relevant when designing multimedia explanations, both in computer-based and book-based environments. These are a multimedia effect, a coherence effect, a spatial contiguity effect, and a personalization effect. The multimedia effect shows that students learn more from words and pictures than from words alone. The coherence effect shows that students learn more when extraneous material is excluded rather than included. The spatial contiguity effect shows that students learn more when corresponding pictures and words are placed near rather than far. The personalization effect shows that students learn more when words are presented in a conversational style than in a formal style. All of these effects work both for computers and books, which shows that the impact of technology is not totally dependent on the nature of the technology but also on the nature of the user. Multimedia systems that capitalize on visual and verbal representations will be more effective in any media environment. As Mayer notes, "media environments do not cause learning; cognitive processing by the learner causes learning (2003, p. 137).

As we discuss more complex technologies, which have a major impact on cognitive processing, it is helpful to keep Mayer's claim in mind. The impact of technology is not independent of the features of the system on which it acts. Thus, taking into consideration universal limitations on cognitive processing, we especially want to draw attention to the way more complex technologies expand human cognitive processing capacities.

Expert Technologies

Expert technologies are those that are scarcely diffused but that are intensively used by a relatively small number of experts. Hutchins and his team (1995; Hutchins & Klausen, 1996) extensively researched the properties of expert technologies. As they note, one of the distinctive features of expert technologies is that they constitute cognitive systems where technologies and their users participate reciprocally. What distinguishes these cognitive systems is that information processing involves the "distribution of a representational state across representational media" (Hutchins & Klausen, 1996, p. 32). Thus, as noted in a famous Hutchins example, information in a plane's cockpit is shared on multiple levels. Each pilot manages some task-relevant information, some information is shared through discourse among the crewmembers, and still other information is located in the physical arrangement of the cockpit. The information moves across these different media of representation. As Hutchins notes, the cognitive interactions among the components of the system are quite complex:

> Certainly, the cognitive properties of the cockpit system are determined in part by the cognitive properties of the individual pilots. They are also determined by the physical properties of the representational media across which a task-relevant representational state is propagated, by the specific organization of the representations supported in those media, by the interactions of metarepresentations held by the members of the crew, and by the distributional characteristics of knowledge and access to task-relevant information across the members of the crew. (Hutchins & Klausen, 1996, pp. 32–33)

Given the close bond between expert technologies and their users, it is appropriate to talk about the performance-enhancing effect of acting *with* these technologies instead of the isolated impact of these technologies. A pilot within a cockpit's technology becomes a system—one that is able to deal with highly complex information through expedited means and can, consequently, make decisions in a dynamic situation. Hoc (this volume) defines dynamic situations as those partially controlled by a human operator. Because events in these situations are not entirely predictable, a great deal of human activity is devoted to the diagnosis of future states. Machines can help a human operator adjust to dynamic situations, but they may also represent an additional source of disturbance. Thus, as Hoc pointed out, human operators need to establish cooperative relationships with these machines, as in the case of distributed representation, previously mentioned. The intellectual impact of expert technologies is, in some ways, similar to what Salomon and Perkins (this volume) call the impact of working *with* a technology. In both cases reference is made to the performance-enhancing impact of artifacts. However, expert technologies have an additional quality. Although some technologies we work *with* can be removed and still have a lasting impact, expert technologies cannot be removed without stopping their effects: They are a substantial part of the cognitive systems they help implement.

To illustrate, consider HAL, the famous computer character in Stanley Kubrick's movie *2001: A Space Odyssey*. HAL, whose role was to take care of the space crewmembers until the end of the mission, experiences an error that puts the mission in peril. In order to ensure the mission's success, the crewmembers decide to deactivate HAL. However, HAL discovers the crew's intentions and kills all save one crewmember, who defeats the homicidal computer in the end. Although Kubrick's computer character is fictional—and anthropomorphic to boot—the saga illustrates the nature of the bond established between expert technologies and their users. Operators in contexts saturated by technology—such as nuclear plants, Antarctic bases, and a plane's cockpit—do indeed depend substantially on the performance of the technology in question. Although conflicts between technologies and operators do not take the personal nature of the conflict between the characters acting in Kubrick's saga, the lack of cooperation between humans and technologies in contexts like those may be, indeed, life threatening (for a review of life conditions in technology rich environments see Harrison, 2002).

Pervasive Technologies

Although infrequently used, some technologies can have a significant impact on human skills. Their impact is not related to their frequency of use but rather to the extent to which they have been disseminated in a given society. Let us illustrate with a paradigmatic example: achievement and aptitude tests. These tests have a pervasive effect on the educational system, as they organize instruction in schools by fostering the skills valued by standardized testing, that is, memory and analytical skills but not creative and practical ones. Instruction can end up being focused on drills for subsequent testing and important educational goals can be subordinated to test preparation (Sternberg, 1996). Nevertheless, tests do not have an effect on education per se, unlike experimental or expert technologies. In fact, there is, arguably, no such thing as an effect of working *with* tests except in those cases where tests are implemented as "dynamic tests" (Sternberg & Grigorenko, 2002a). Dynamic tests evaluate a student's learning potential and, at the same time, help to increase the skills they assess. Conversely, traditional tests are focused on the static assessment of skills acquired before the testing session. As is widely known, test drills help students achieve better scores on standardized tests. Test drills are, of course, motivated by the fact that tests are gatekeepers for greater educational opportunities. Thus, traditional tests have an impact on education because of their role in socially meaningful activities, notably, opening or closing the doors to future educational and work opportunities. This is the main reason why their impact is so pervasive. The same applies to other technologies (such as those related to the application of the law, for instance).

Thus, despite their impact on everyday life, pervasive technologies are not the most influential category of technology, as their impact is derived from, and is secondary to, their social nature.

Radical Technologies

We define highly disseminated and frequently used technologies as radical. We characterize them as radical because they have the power to restructure human cognition. They do not have isolated effects on specific skills but shape the way information is processed and represented more fundamentally. Thus, effects of radical technologies correspond more or less to what

Salomon and Perkins (this volume) call effects *through* technology; that is to say, their impact reorganizes cognitive activity.

To clarify the nature of radical technologies, we focus on two that form the core of the schooling process: written scripts and numerals. Because of their power to restructure intellectual activity, written scripts and numerals have been called "the major tools of thought" (Bruner, 1966, p. 112). As such, competence with linguistic and numeric symbols is one of the primary targets of elementary education. Spreading literacy has been promoted by international agencies such as UNESCO (Olson, 1994). Additionally, a number of international studies such as the Programme for International Student Assessment (PISA) (Adams & Wu, 2002) and the Trends in International Mathematics and Science (TIMMS) (Mullis et al., 2000) studies consider students' mathematical performance as a fundamental index of a country's educational efficacy. Moreover, linguistic and mathematical skills have traditionally constituted one of the privileged ways to test "intelligence." Indeed, intelligence tests frequently include items based on reading comprehension, vocabulary, and mathematical problem solving, among other things (Mackintosh, 1998).

Today, a good number of scholars see aptitudes for language and mathematics as part of the collection of natural endowments of the human mind. Thus, blending Darwin (1859) and Chomsky (1975), many researchers have adopted the view that the human brain has an evolved predisposition toward language and numeracy, among other capacities (Pinker, 1997; Pinker & Bloom, 1990; Wynn, 1992). These scholars focus a significant part of their work on infants and preschoolers, as they want to minimize the impact of culture as much as possible. Complementing these claims, we consider written language and numerals as exemplary cases of the impact culture has on cognitive development. As has been noted elsewhere (Tomasello, 1999a), systems of representation, such as written language and mathematical numerals, engage the accumulation and transmission of cultural resources by means of social interaction. Language (written language in particular) and mathematics (numerals in particular) are not simply evolved dispositions but major historical inventions. As Nickerson (this volume) notes:

> Presumably as soon as humans learned to count and to measure, they made devices to help them to do so and to remember the results. The development of symbol systems and written language was certainly among the most noteworthy technological achievements of prehistory; there is no other technological advance whose effects on human history rival those of this one.

Olson (1994; 1996; this volume) proposes that written scripts have a singular impact on the way we approach language as well as a distinctive impact on intelligence. From his point of view, written scripts are more than transcribed speech. In their historical evolution, written scripts polished a particular syntax and became models of speech. Once a script crystallizes as an accomplished system of representation, it "provides the model, a set of distinctive but related concepts and categories however distorted and fragmentary, in terms of which one can analyze and so become aware of certain basic properties of one's speech" (Olson, 1996, p. 86). The developmental side of this feature of scripts is that learning a script, particularly an alphabetic one, provides a new kind of metalinguistic awareness. "In learning to read and write one is learning not only a skill but learning to think about language and mind in a new way. This learning is summarized in the concept of metarepresentation. Whereas language is about, and in that sense represents, the world, writing is a representation of language, hence, a metarepresentation" (Olson, this volume). For Olson, this metarepresentation involves a consciousness of the higher-order features of language, that is, words, sentences, and the documentary practices that are born from them, such as creating dictionaries. Concerning the relationship between writing and intelligence, Olson suggests that this metalinguistic awareness provides intelligence with its basic character. Intelligence tests are tests of our competence as literate beings. In fact, IQ tests include items that deal with vocabulary and the relationship between words, thus testing our literate capacity.

Nevertheless, the impact of written language on cognition is not a consequence of the nature of a script per se. Its impact derives from the use a script in everyday life. This interpretation was supported by a study developed in Liberia by Scribner and Cole (1981). Their main goal was to test the *literacy hypothesis* in its strong version, that is, the hypothesis that the acquisition of a script has substantial cognitive consequences, which were imagined to be a precursor of the development of abstract thought (Olson, 1994). Their second goal was to test whether the impact of literacy was part and parcel of the impact of schooling, or whether it reflected proficiency with a specific script (Cole, 1996). In Liberia, Scribner and Cole (Cole, 1996; Scribner & Cole, 1981) found a unique opportunity to explore these issues, particularly among the Vai people. Some of the Vai were literate in a script without being schooled and some of them were fluent in three different scripts: 20% of the people they studied were literate in Vai, 16% were partially literate in Arabic, and about 6% had learned English in school. Exploring the cognitive skills of these three groups, Scribner and Cole were able to show effects that were script-specific. For instance, the Vai script represents language syllabically and Vai literates outperformed nonliterates on a task that involved analyzing language at a syllabic level. However, those individuals exposed to Qur'anic classes, where people learned to memorize in an incremental fashion, were especially good at tasks that involved memorizing a list of items whose length increased by one item per trial, that is, incrementally. Most important, Scribner and Cole showed that schooling, and not a specific script, was most likely to improve performance in a variety of cognitive tasks involving categorizing, memory, logical reasoning, encoding and decoding, semantic integration, and verbal explanation. In addition, they concluded that it is not the knowledge of a script per se that matters, but rather its uses in everyday life. As Cole (1996) noted, "if the uses of writing are few, the skill development they foster will also be limited to a narrow range of tasks in a correspondingly narrow range of activities and content domains" (p. 235). Thus, the larger the diffusion of a writing system within a context, the more pervasive are its cognitive effects.

Let us now comment on the corresponding cognitive consequences of written numerals. Zhang & Norman (1994) proposed that written numerals allow for the distribution of the operation of mathematical calculations between external representations and internal representations. External representations make a mathematical task easier because they activate perceptual processes that complement the mnemonic processes triggered by internal representations. Different representations of numbers allow different kinds of mathematical performance. That is, the mind does not perform identical operations when using an abacus or using Roman numerals to add. Therefore, although all numeric systems represent numbers, they differ in their levels of relative efficiency. Even though the Arabic system is not necessarily the more efficient (according to the authors, the abacus is more efficient), it became the dominant system of representation for numerical calculations for a number of reasons: "It integrates representation and calculation into a single system, in addition to its other nice features of efficient information encoding, compactness, extendibility, spatial representation, small base, effectiveness of calculation and, especially important, ease of writing" (Zhang & Norman, 1994, p. 293). Tomasello (1999a) also called attention to the relative superiority of Arabic numerals. He proposes that "the Arabic system of enumeration is much superior to older Western systems for purposes of complex mathematics (e.g., Roman numerals), and the use of Arabic numerals, including zero and the place system for indicating different-sized units, opened up for Western scientists and other persons whole new vistas of mathematical operations" (pp. 45–46).

Still, as happens with written scripts, the impact of numerals is also mediated by their everyday relevance. The canonical systems of representation—that is, the alphabet or Arabic numerals—are not always the best available tools for the adaptive needs of an individual. People can display adaptive mathematical performance without making use of Arabic numerals. Research on "working intelligence" (Scribner, 1986) or "practical intelligence" (Sternberg et al., 2000; Sternberg & Wagner, 1986; Wagner & Sternberg, 1985) shows that there are plenty of cases where people perform better on real-life mathematical problems than on comparable problems formulated using standard scholarly mathematics. To illustrate, let us reference some classic studies in practical mathematics by Scribner (1986), Saxe (1989/1997), Lave (Lave, Murtaugh, & de la Rocha, 1984/2000), and Ceci (Ceci & Liker, 1987; Ceci & Roazzi, 1994). Scribner (1986) showed that New York cargo loaders involved in physical tasks were able to reduce their mental and physical workload by calculating, based on the symbols available at their work environment, the least number of physical moves required to perform their task. By decoding the symbols that were on hand, the loaders translated a number of mathematical operations into a set of particular physical maneuvers that minimized their physical load. Moreover, when mathematical problems were presented in their context of work, the loaders performed better than control groups possessing a higher level of academic training. Thus, whatever made their performance efficient was not their level of mathematical skill as measured by conventional mathematical problems but their mathematical skill as measured by a set of symbols related to their practical working needs. Scribner's studies also showed that intellectual work and manual labor should not be treated differently. Tasks that are usually labeled as manual or blue-collar often involve complex intellectual performance. Saxe (1989/1997) produced similar findings when studying candy sellers. He showed that 10- to 12-year-old Brazilian street vendors were able to perform quite complex mathematical operations that did not correlate with their performance in standard mathematical tests. He found that children "had developed an ability to use bills themselves as signifiers for large values and did not need to rely on their imperfect knowledge of the standard number orthography" (Saxe, 1989/1997, p. 335). These operations were quite adaptive, as the Brazilian economy was affected by a high inflation rate, so the children's ability to perform exact price calculations was key for their survival.

Ceci and colleagues found similar results in different samples of expert racetrack handicappers and street children. Handicappers' ability to predict post-time odds at the racetracks was unrelated to their IQ. Brazilian street children's ability to perform appropriate mathematical calculations in everyday relevant tasks was similarly unrelated to their performance in formal mathematics. Finally, Lave and collaborators (1984/2000) reported congruent results for the activity of grocery shopping. The ability of shoppers to obtain the best value was unrelated to performance in the M.I.T. test of mental arithmetic. In short, people can act mathematically by making use of symbols that are context specific.

Before closing this section let us make a few more points about the affordances of radical technologies. Norman (1988) notes that tools have specific *affordances*, a term originally coined by James Gibson (1977). For Gibson, affordances describe the reciprocal relationship between the world and a person or animal. They are resources the environment offers the animal, which must have the capability to perceive them. (For instance, surfaces provide support.) Some affordances are detectable; others are not (E. J. Gibson, 1999). Norman applies the term to artifacts and notes that they afford the user the ability to execute certain behaviors to achieve his or her practical goals. For Norman, good design involves creating artifacts whose affordances are easy to perceive and to understand—so-called user-friendly tools. When employing user-friendly tools, we are more or less aware of their affordances. In contrast, when operating with radical technologies such as numerals or a script, we are barely conscious of what they afford. In fact, once a written or a numeric system is acquired, it becomes more or less invisible. Thus, typical Western individuals who use an alphabet and Arabic numerals can hardly look at words and quantities without the lens provided by the alphabet and the Arabic numerals. Concerning literate language, Olson noted, "the general finding is that people familiar with an alphabet come to *hear* words as composed of the sounds represented by the letters of the alphabet: those not familiar do not" (1996, p. 93). In short, once tools are internalized, their cultured use confirms the metaphor that says we are like fish in the water of culture (Tomasello, 1999a).

A REVISED DEFINITION OF INTELLIGENCE

During the 20th century, a restless group of psychologists promulgated a notion of intelligence as general, predominantly innate, and rooted in elementary processes. This view originated in the work of Anglo-Saxon theoreticians such as Galton (1883) and Spearman (1904; 1927). It grounded the theory of the general factor of intelligence, that is, the idea that a single factor of intelligence accounts for the positive correlation between different kinds of scholastic and psychometric tasks (for a historical review see Sternberg, 1990). Today the "*g*-ocentric" view is ubiquitous among experts—and among the general public as well, in part through the best-selling book *The Bell Curve* (Hernstein & Murray, 1994). Moreover, a number of contemporary researchers claim a central role for *g* in the study of human intelligence (see essays in Sternberg & Grigorenko, 2002b). For instance, famous *g* theorist Arthur Jensen claims:

> The construct known as psychometric *g* is arguably the most important construct in all psychology largely because of its ubiquitous presence in all tests of mental ability and its wide-ranging predictive validity for a great many socially significant variables, including scholastic performance and intellectual attainments, occupational status, job performance, income, law abidingness, and welfare dependency. (2002 p. 39)

We believe that a consideration of technology transcends theories of intelligence based on the *g*-factor. As we explain, our consideration of technology portrays human intelligence as dynamic, context-dependent, culturally shaped, multiple, and distributed.

A View of Intelligence as Dynamic

Most *g*-theories are static theories of intelligence: They conceive of intelligence as a genetically endowed property of the mind that saturates all cognitive tasks a person performs, as demonstrated by factor analysis of cognitive tasks (Sternberg & Grigorenko, 2002b). When cognitive tools are taken into consideration, the image of intelligence that arises is quite distinct from *g*-theory. Indeed, a consideration of technology drives us to see intelligence as shaped by the external resources an individual has on hand: a script, a numerical system, a map, or a computer, just to mention a few. As Nickerson (this volume) highlights, these cognitive tools extend the abilities individuals have to deal with the challenges of everyday life.

Cognitive tools are not static. They are historical inventions and can be improved through a process of selection acting at a cultural level. In turn, their intellectual effects are relative to this process. Cumulative cultural evolution depends on two basic processes, innovation and imitation, which are supplemented by instruction (Kruger & Tomasello, 1996; Tomasello, 1999a, 1999b, 2000; Tomasello, Kruger, & Ratner, 1993). A generation invents an artifact, which is passed to a second generation via imitation and instruction. The second generation modifies and passes the modified artifact to a third generation, who continue the cycle of imitation and innovation. And so it goes. Some technologies remain and significantly mold human cognition, such as literate technologies; others fade away, such as slide rules did at the end of the 20th century after the invention of pocket calculators (Nickerson, this volume). In consequence, the impact of technology on human abilities is not static. On one hand, technology shapes human skills. On the other hand, technologies are shaped through cultural evolution. And technology evolves as the problems it addresses change as well. Some tools are replaced as more efficient ones are invented; others fade away as the tasks they helped to solve become obsolete. Thus, given its relative dependence on the dynamism of cultural evolution, intelligence is not static either.

Cumulative cultural evolution proceeds not only between generations but also between contexts. In effect, transfers of technology from the developed world to the developing world reproduce cultural evolution, but they do so synchronically. The users adapt the imported technology to their local realities through a process of innovation, which allows for the assimilation and accommodation of the imported tools. Although the providers of technology do not usually take psychological processes into consideration, the local users develop myriad solutions to adjust the tools to their geographical and climatic context as well as to their level of expertise. Thus, the transfer of technology produces multiple possible work arrangements and multiple ways of distributing cognition between users and tools. However, technological transfers do not necessarily imply progress. The exceptional and successful cases are those that capitalize on the creative potential of the receiving population in order to introduce changes in the design of those technologies accordingly (Díaz-Canepa, this volume).

A View of Intelligence as Culturally Shaped

There is a phenomenon that neatly illustrates the way culture shapes human intelligence and that is strongly related to tool use: the continuous rise of IQ scores of large cross-sectional samples of test takers from developed nations across the 20th century (Flynn, 1987). Known as the "Flynn effect," after its discoverer, James Flynn, this phenomenon is a telling rebuttal to the predictions made by early 20th century eugenicists, who predicted a decrease of the "intelligence of the nations" because of immigration or racial blending (Neisser, 1998). The Flynn effect not only shows that those predictions were mistaken but that the opposite phenomenon occurred: IQ scores increased, and they did so in what is supposedly the more genetically driven area of IQ—abstract-thinking and visual-thinking skills. Greenfield suggests that these increases are caused, in part, by the diffusion of technologies that make extensive use of visual skills. Greenfield and collaborators have presented evidence that computer use indeed improves visual and spatial skills (Subrahmanyam, Greenfield, Kraut, & Gross, 2001), a finding that has been replicated by other researchers as well (Okagaki & Frensch, 1994). In particular, the evidence shows that expertise in computer games is related to improvements in attention, the development of iconic and spatial representations, and improved performance on tasks involving mental transformations (Maynard, Subrahmanyam, & Greenfield, this volume).

Thus, as noted by Kirlik (this volume), the study of technology puts a real face on what psychology usually treats as an undifferentiated variable: environment. Complementing Flynn's findings, research has shown that environment does not affect IQ equally across "contexts." For instance, the "heritability" of intelligence radically changes across socioeconomic groups. In impoverished families, 60% of the variance in IQ is accounted for by shared environment; in affluent families the results are the opposite (Turkheimer, Haley, Waldron, D'onofrio, & Gottesman, 2003). The researchers note that "the developmental forces at work in poor environments are qualitatively different from those at work in adequate ones" (Turkheimer et al., 2003, p. 6). One plausible explanation for these differences may be related to technology use. It is valid to hypothesize that affluent environments may provide a developing child with more cognitive tools than impoverished environments. Because of the favorable impact of cognitive tools on intellectual development, genetic differences are accentuated, as developing children actualize more of their potential, a hypothesis aligned with the bioecological model drawn by Bronfrenbrenner and Ceci (1994).

A Multiple View of Intelligence

Our consideration of technology puts an emphasis on the distinctive skills that constitute intelligence. Therefore, from the perspective of tool use, intelligence is not seen as unitary, but as consisting of a set of context-relevant skills. First, at the design level, technologies target specific skills. We discussed previously, for instance, the case of the Reading Partner, which targets reading and writing skills. Many other cases of computer software exist whose content is relatively domain specific. Second, at the information-processing level, technologies capitalize on the main channels humans have to process information. For instance, as previously mentioned, it has been proposed that multimedia technologies must take into consideration the two channels (visual and verbal) through which humans process information. Third, at the level of intellectual training, more widespread technologies foster specific skills. As we commented, computer games favor the development of visual–spatial thinking and may be one of the factors underlying the Flynn effect. However, computer games do not as much favor the verbal component of intelligence, as evaluated by tests of crystallized intelligence (Greenfield, 1998; Subrahmanyam et al., 2001).

The technology-oriented view of intelligence shares with some recent evolutionary psychology approaches (Barkow, Cosmides, & Tooby, 1992) a model of the mind as composed of a diverse set of skills. However, these perspectives also have differences that are worth noting. First, the technological point of view emphasizes the distribution of information processing between human and artifact, whereas the evolutionary view rests on hypothetical conceptual domains, such as biology or physics, which organize information processing internally (Hirschfeld & Gelman, 1994). Second, whereas the technological point of view awards culture a central role in the shaping of the human mind, evolutionary psychology sees culture as an auxiliary influence (Tooby & Cosmides, 1992). For similar reasons, the technological view may be seen as compatible with Gardner's theory of multiple intelligences (Gardner, 1999) as well. Gardner's theory acknowledges the role played by cognitive tools in the context of what he calls symbol systems. As he states, "The human brain seems to have evolved to process certain kinds of symbols efficiently. Put differently, symbol systems may have been developed precisely because of their preexisting, ready fit with the relevant intelligence or intelligences" (1999, p. 38). Unlike Gardner's model, our view of cognitive tools is not dependent on an evolutionary story, but rather, contingent on the creative potential of individuals to produce new technologies and, in turn, to be shaped by them.

A View of Intelligence as Distributed

A psychological consideration of technology is usually associated with work on distributed cognition (Salomon, 1993). By distributed cognition, we refer to the fact that the representation and processing of information are distributed between a person and the artifacts this person uses. Whereas some scholars adopt a strong stance and think that the whole idea of cognition should be reconceived to take technology into consideration, others take a relatively weaker stance and make a distinction between human information processing and distributed processing (Salomon, 1993).

Pea (1993) called attention to the fact that there are two forms of distributed intelligence. One form is the co-construction of intelligence in joint actions such as in parent–child or classroom interaction, which he calls "social distribution of intelligence" (p. 50). The other form is created through the goal-oriented use of parts of the environment or of artifacts, which he terms "material distribution of intelligence" (p. 50). We deem both as interdependent forms of cognition. Artifacts have a more substantial impact when they are anchored in socially relevant practices (Cole, 1996; Preiss & Sternberg, 2003). Yet Pea's distinction is useful, as it makes evident the fact that artifacts crystallize the intellectual practices of a community. As Pea writes:

> These tools literally carry intelligence in them, in that they represent some individual's or some community's decision that the means thus offered should be reified, made stable, as a quasi-permanent form, for use by others. In terms of cultural history, these tools and the practices of the user community that accompany them are major carriers of patterns of previous reasoning. (1993, p. 53)

The diffusion of technologies in the workplace has typically been related to the fear of de-skilling and losing a job, a fear once felt predominantly by people working on traditional lines of production, but now increasingly experienced by so-called white- collar workers. As the technology they manage is replaced by complex work systems, some of their job positions become obsolete. However, these changes, in fact, trigger not de-skilling, but instead more innovative means of skill acquisition. Indeed, the new work systems will need individuals with a great deal of cognitive flexibility. For instance, it has been suggested that advanced manufacturing in newly industrialized countries such as Mexico, depends "on the effectiveness with which workers could acquire new skills, especially the ability to maintain and quickly repair complex equipment" (Shaiken, 1998, p. 279). As most workers in emerging economies are not suitably trained, they have to rapidly develop an appropriate level of expertise. According to Shaiken, the most efficient way to do so is through teamwork. Thus, although complex cognitive skills have, in fact, been quite resilient in the face of technological substitution, contemporary technologies might foster their development.

CONCLUDING WORDS

A review of recent literature on the interaction between intelligence and technology evidences a growing convergence. There is an increasing awareness not only of the way technology shapes human activity but also of the way human cognition is shaped by its evolved capacity to profit from cultural tools. The research on technology reviewed in this chapter resituates culture as the focal point of the study of intelligence. It has recently been suggested that the mind is not a blank slate because its evolved dispositions frame the ways information is processed (Pinker, 2002). Indeed, the human mind has evolved a capacity to make use of cultural tools (Bruner, 2002). Research on intelligence and technology shows that culture is not a blank slate either. Cultural tools are invented historically and transmitted from one generation to the next and acquired ontogenetically. Some tools that are commonplace to one generation were created only through a great intellectual struggle by the previous generation. As these tools become commonplace and shared by a larger group of people, cognition becomes increasingly technological. As Pea notes, "the inventions of Leibniz's calculus and Descartes's coordinate graphs were startling achievements; today they are routine content for high school mathematics" (1993, p. 53).

Culture is relevant for several aspects of human intelligence. First, culture constitutes the background for ontogenetic development. As Hutchins writes, "symbols are in the world first, and only later in the head" (1995, p. 370). Enculturation involves the internalization of symbols and tools. Second, culture configures the nature of a situation in which an individual acts. Third, culture stores what we have done in the world and allows for the transmission of articulated information from one generation to the next. That is to say, culture makes it possible for our creative inventions to have lasting effects.

It has been proposed elsewhere that culture-free artifacts do not exist (Cole, 1996). Artifact-free abilities do not exist either. "Each form of experience, including the various symbolic systems tied to the media, produces a unique pattern of skills for dealing with or thinking about the world. It is the skills in these systems that we call intelligence" (Olson & Bruner, 1974, p. 149). The mastery of an artifact involves the learning of a specific skill, but it also entails a meaningful expansion of our intellectual capabilities. Therefore, what is finally acquired is not a universal skill but *a skill that is intrinsically connected to an artifact.* Through their socialization in different artifacts, people continuously reshape their intelligence. In doing so, they build a set of skills that are not only context relevant but also culturally nurtured.

ACKNOWLEDGMENTS

During manuscript preparation, the first author was supported by grants from the Fulbright Commision, Yale University, and the Chilean Government. He expresses his gratitude. The first author thanks Josephine Fueser and Steven Shafer for help with manuscript preparation.

REFERENCES

Adams, R., & Wu, M. (Eds.). (2002). *PISA 2000 Technical Report.* Paris: Organisation for Economic Co-Operation and Development.

Barkow, J. H., Cosmides, L., & Tooby, J. (Eds.). (1992). *The adapted mind. Evolutionary psychology and the generation of culture.* New York: Oxford University Press.

Bogin, B. (2001). *The growth of humanity.* New York: Wiley-Liss.

Boysen, S. T., & Hallberg, K. I. (2000). Primate numerical competence: Contributions toward understanding nonhuman cognition. *Cognitive Science, 24*(3), 423–443.

Bronfrenbrenner, U., & Ceci, S. (1994). Nature and nurture reconceptualized. A biological model. *Psychological Review, 101*(4), 568–586.

Bruner, J. S. (1966). *Towards a theory of instruction.* Cambridge, MA: Harvard University Press.

Bruner, J. S. (2002). *Making stories: Law, literature, life.* New York: Farrar, Straus & Giroux.

Ceci, S. J., & Liker, J. K. (1987). A day at the races: A study of IQ, expertise, and cognitive complexity. *Journal of Experimental Psychology: General, 116*(2), 90.

Ceci, S. J., & Roazzi, A. (1994). The effects of context on cognition: Postcards from Brazil. In R. J. Sternberg (Ed.), *Mind in context: Interactionist perspectives on human intelligence* (pp. 74–101). New York: Cambridge University Press.

Chomsky, N. (1975). *Reflections on language.* New York: Random House.

Cole, M. (1996). *Cultural psychology: A once and future discipline.* Cambridge, MA: Harvard University Press.

Darwin, C. (1859). *On the origin of species.* London: Murray.

Ellul, J. (1980). *The technological system.* New York: Continuum.

Encyclopædia Britannica Online. (2004a). *Technology*. Retrieved June 11, 2004, from http://search.eb.com/

Encyclopædia Britannica Online. (2004b). *Tool*. Retrieved June 11, 2004, from http://search.eb.com/

Flynn, J. R. (1987). Massive IQ gains in 14 nations: What IQ tests really measure. *Psychological Bulletin, 101*(2), 171–191.

Galton, F. (1883). *Inquiry into human faculty and its development*. London: Macmillan.

Gardner, H. (1999). *Intelligence reframed: Multiple intelligences for the 21st century*. New York: Basic Books.

Gibson, E. J. (1999). Affordances. In R. Wilson & F. Keil (Eds.), *The MIT encyclopedia of the cognitive sciences* (pp. 4–6). Cambridge, MA: The MIT Press.

Gibson, J. J. (1977). The theory of affordances. In R. E. Shaw & J. Bransford (Eds.), *Perceiving, acting, and knowing* (pp. 67–82). Hillsdale, NJ: Lawrence Erlbaum Associates.

Goodenough, W. H. (1994). Toward a working theory of culture. In R. Borovsky (Ed.), *Assessing cultural anthropology* (pp. 262–273). New York: McGraw-Hill.

Greenfield, P. M. (1998). The cultural evolution of IQ. In U. Neisser (Ed.), *The rising curve: Long-term gains in IQ and related measures* (pp. 81–125). Washington, DC: American Psychological Association.

Grigorenko, E. L., Jarvin, L., Niu, W., & Preiss, D. (in press). Is there a standard for standardized testing? In L. Stankov (Ed.), *Extending intelligence: Enhancement and new constructs*. Mahwah, NJ: Lawrence Erlbaum Associates.

Hanakawa, T. M. H., Okada, T., Fukuyama, H., & Shibasaki, H. (2003). Neural correlates underlying mental calculations in abacus experts: A functional magnetic resonance imaging study. *NeuorImage, 19*, 296–307.

Harrison, A. (2002). *Spacefaring. The human dimension*. Los Angeles: University of California Press.

Hatano, G. (1997). Learning arithmetic with an abacus. In P. Bryant (Ed.), *Learning and teaching mathematics: An international perspective* (pp. 209–232). Hove, UK: Psychology Press.

Hernstein, R. J., & Murray, C. (1994). *The bell curve*. New York: Free Press.

Hirschfeld, L. A., & Gelman, S. A. (Eds.). (1994). *Mapping the mind. Domain specificity in cognition and culture*. New York: Cambridge University Press.

Hutchins, E. (1995). *Cognition in the wild*. Cambridge, MA: The MIT Press.

Hutchins, E., & Klausen, T. (1996). Distributed cognition in an airline cockpit. In D. Middleton (Ed.), *Cognition and communication at work* (pp. 15–34). New York: Cambridge University Press.

Ifrah, G. (2000). *The universal history of computing*. New York: Wiley.

Jensen, A. R. (2002). Psychometric *g*: Definition and substantiation. In E. Grigorenko (Ed.), *The general factor of intelligence. How general is it?* (pp. 39–55). Mahwah, NJ: Lawrence Erlbaum Associates.

Kruger, A. C., & Tomasello, M. (1996). Cultural learning and learning culture. In N. Torrance (Ed.), *The handbook of education and human development: New models of learning, teaching and schooling* (pp. 368–387). Cambridge, MA: Blackwell.

Lajoie, S. (2000). *Computers as cognitive tools, volume two: No more walls*. Mahwah, NJ: Lawrence Erlbaum Associates.

Lajoie, S., & Derry, S. J. (1993). *Computers as cognitive tools*. Hillsdale, NJ: Lawrence Erlbaum Associates.

Latour, B. (1993). *We never have been modern*. Cambridge, MA: Harvard University Press.

Lave, J., Murtaugh, M., & de la Rocha, O. (2000). The dialectic of arithmetic in grocery shopping. In J. Lave (Ed.), *Everyday cognition* (pp. 67–94). Bridgewater, NJ: Replica Books. (Original work published 1984)

Lee, C. D. (2003). Toward a framework for culturally responsive design in multimedia computer environments: Cultural modeling as a case. *Mind, Culture and Activity, 10*(1), 42–61.

Mackintosh, N. J. (1998). *IQ and human intelligence.* New York: Oxford University Press.

Mayer, R. E. (2003). The promise of multimedia learning: Using the same instructional design methods across different media. *Learning and Instruction, 13,* 125–139.

Mullis, I. V. S., Martin, M. O., Gonzalez, E. J., Gregory, K. D., Garden, R. A., O'Connor, K. M., Chrostousti, S. J., & Smith, T. A. (2000). *TIMSS 1999 international mathematics report. Findings from IEA's repeat of the Third International Mathematics and Science Study at the eighth grade.* International Study Center, Lynch School of Education, Boston College, Massachusetts.

Musallam, S., Corneil, B. D., Greger, B., Scherberger, H., & Andersen, R. A. (2004). Cognitive control signals for neural prosthetics. *Science, 305,* 258–262.

Neisser, U. (1976). General, academic, and artificial intelligence. In L. Resnick (Ed.), *Human intelligence: Perspectives on its theory and measurement* (pp. 179–189). Norwood, NJ: Ablex.

Neisser, U. (1998). Introduction: Rising test scores and what they mean. In U. Neisser (Ed.), *The rising curve: Long-term gains in IQ and related measures* (pp. 3–22). Washington, DC: American Psychological Association.

Norman, D. A. (1988). *The psychology of everyday things.* New York: Basic Books.

Okagaki, L., & Frensch, P. A. (1994). Effects of video game playing on measures of spatial performance: Gender effects in late adolescence. *Journal of Applied Developmental Psychology, 15*(1), 33–58.

Olson, D. R. (1994). *The world on paper: The conceptual and cognitive implications of writing and reading.* New York: Cambridge University Press.

Olson, D. R. (1996). Towards a psychology of literacy: On the relations between speech and writing. *Cognition, 60*(1), 83–104.

Olson, D. R., & Bruner, J. S. (1974). Learning through experience and learning through media. In D. R. Olson (Ed.), *Media and symbols: The forms of expression, communication, and education* (pp. 125–150). Chicago: University of Chicago Press.

Pea, R. D. (1993). Practices of distributed intelligence and designs for education. In G. Salomon (Ed.), *Distributed cognitions. Psychological and educational considerations* (pp. 47–87). New York: Cambridge University Press.

Pinker, S. (1997). *How the mind works.* New York: Norton.

Pinker, S. (2002). *The blank slate: The modern denial of human nature.* New York: Viking.

Pinker, S., & Bloom, P. (1990). Natural language and natural selection. *Behavioral & Brain Sciences, 13*(4), 707–784.

Plotkin, H. (2003). *The imagined world made real: Towards a natural science of culture.* New Brunswick, NJ: Rutgers University Press.

Preiss, D., & Sternberg, R. J. (2003). Prácticas Intelectuales en el Trabajo: Conocimiento, Actividad y Tecnología [Intellectual practices at work: Knowledge, activity and technology]. *Psykhe: Revista de la Escuela de Psicologia, 11*(2), 3–16.

Quartz, S. R., & Sejnowski, T. J. (2002). *Liars, lovers, and heroes.* New York: Morrow.

Reinking, D. (Ed.). (1998). *Handbook of literacy and technology: Transformations in a post-typographic world.* Mahwah, NJ: Lawrence Erlbaum Associates.

Salomon, G. (Ed.). (1993). *Distributed cognitions. Psychological and educational considerations.* New York: Cambridge University Press.

Salomon, G., Globerson, T., & Guterman, E. (1989). The computer as a zone of proximal development: Internalizing reading-related metacognitions from a reading partner. *Journal of Educational Psychology, 81*(4), 620–627.

Saxe, G. B. (1997). Selling candy: A study of cognition in context. In O. Vasquez (Ed.), *Mind, culture and activity. Seminal papers from the Laboratory of Comparative Human Cognition* (pp. 330–337). Cambridge, UK: Cambridge University Press. (original work published 1989)

Scribner, S. (1986). Thinking in action: Some characteristics of practical thought. In R. Wagner (Ed.), *Practical intelligence: Origins of competence in the everyday world* (pp. 143–162). New York: Cambridge University Press.

Scribner, S., & Cole, M. (1981). *The psychology of literacy.* Cambridge, MA: Harvard University Press.

Semaw, S., Rogers, M. J., Quade, J., Renne, P. R., Butler, R. F., Dominguez-Rodrigo, M., Stout, D., Hart, W. S., Pickering, T. R., & Simpson, S. W. (2003). 2.6-million-year-old stone tools and associated bones from OGS-6 and OGS-7, Gona, Afar, Ethiopia. *Journal of Human Evolution, 45*(2), 169–177.

Shaiken, H. (1998). Experience and the collective nature of the skill. In D. Middleton (Ed.), *Cognition and communication at work* (pp. 279–296). New York: Cambridge University Press.

Spearman, C. (1904). General intelligence, objectively determined and measured. *American Journal of Psychology, 15*(2), 201–293.

Spearman, C. (1927). *The abilities of man: Their nature and measurement.* New York: Macmillan.

Sternberg, R. J. (1990). *Metaphors of mind: Conceptions of the nature of intelligence.* New York: Cambridge University Press.

Sternberg, R. J. (1996). *Successful intelligence: How practical and creative intelligence determine success in life.* New York: Simon & Schuster.

Sternberg, R. J., Forsythe, G. B., Hedlund, J., Horvath, J. A., Wagner, R. K., Williams, W. M., Snook, S., & Grigorenko, E. L. (2000). *Practical intelligence in everyday life.* Cambridge, UK: Cambridge University Press.

Sternberg, R. J., & Grigorenko, E. (2002a). *Dynamic testing: The nature and measurement of learning potential.* Cambridge, UK: Cambridge University Press.

Sternberg, R. J., & Grigorenko, E. (Eds.). (2002b). *The general factor of intelligence. How general is it?* Mahwah, NJ: Lawrence Erlbaum Associates.

Sternberg, R. J., & Wagner, R. K. (1986). *Practical intelligence: Nature and origins of competence in the everyday world.* New York: Cambridge University Press.

Subrahmanyam, K., Greenfield, P., Kraut, R., & Gross, E. (2001). The impact of computer use on children's and adolescents' development. *Journal of Applied Developmental Psychology, 22*(1), 7–30.

Tanaka, S., Michimata, C., Kaminaga, T., Honda, M., & Sadato, N. (2002). Superior digit memory of abacus experts: An event-related functional MRI study. *NeuroReport, 13*(17), 2187–2191.

Tomasello, M. (1999a). *The cultural origins of human cognition.* Cambridge, MA: Harvard University Press.

Tomasello, M. (1999b). The human adaptation for culture. *Annual Review of Anthropology, 28,* 509–529.

Tomasello, M. (2000). Culture and cognitive development. *Current Directions in Psychological Science, 9*(2), 37–40.

Tomasello, M., Kruger, A. C., & Ratner, H. H. (1993). Cultural learning. *Behavioral and Brain Sciences, 16,* 495–552.

Tooby, J., & Cosmides, L. (1992). Introduction: Evolutionary psychology and conceptual integration. In J. Tooby (Ed.), *The adapted mind: Evolutionary psychology and the generation of culture* (pp. 19–136). New York: Oxford University Press.

Turkheimer, E., Haley, A., Waldron, M., D'onofrio, B., & Gottesman, I. I. (2003). Socioeconomic status modifies heritability of IQ in young children. *Psychological Science, 14*(6), 623–628.

Urton, G. (2003). *Quipu: Contar anudando en el imperio Inka* [Quipu: Knotting account in the Inca Empire]. Santiago, Chile: Museo Chileno de Arte Precolombino & Harvard University.

Valsiner, J., & van der Veer, R. (2000). *The social mind: Construction of the idea.* Cambridge, UK: Cambridge University Press.

Vygotsky, L. S. (1978). *Mind in society: The development of higher psychological processes.* Cambridge, MA: Harvard University Press.

Wagner, R. K., & Sternberg, R. J. (1985). Practical intelligence in real-world pursuits: The role of tacit knowledge. *Journal of Personality & Social Psychology, 49*(2), 436–458.

Wynn, K. (1992). Addition and subtraction by human infants. *Nature, 358*(6389), 749–750.

Zhang, J., & Norman, D. A. (1994). A representational analysis of numeration systems. *Cognition, 57*(3), 271–295.

10

We Have Met Technology and It Is Us

Michael Cole
University of California, San Diego

Jan Derry
University of London

Stanley Kubrick's film *2001 A Space Odyssey* provides a useful tool for thinking about the relationship between intelligence and technology. The opening scenes depict a prehistoric time when a troop of ape-like creatures inhabit an environment that is similar to contemporary images of the African savanna. The ape-like creatures forage side by side with pig-like animals in peaceful harmony, and both the apes and the pigs are prey to large members of the feline persuasion. The apes quarrel with each other and with other bands about access to a watering hole during which they jump up and down, making threatening sounds and gestures, but never directly kill their competitors. At night they huddle together beneath rock ledges, listening fearfully to the noises of dangerous predators.

One morning they awake to find a giant, black, steel rectangle lodged before them. It is clearly not a part of the natural world they have inhabited up to that time. As the sun rises over this gigantic object (not unlike a huge domino made entirely of black Teflon of the kind you see at the bottom of a modern frying pan), one of the apes picks up the leg bone of an animal that has been killed in some previous encounter with a killer-cat, and the image of a pig dying appears on the screen—an anticipatory representation in the dawning mind of the hungry ape. "Thought" turns into action. Now instead of peacefully grazing alongside the pigs, the bleached femur of another animal serves as a means of killing them, and instead of jumping up and down and threatening competing bands of apes at a water hole, an alpha ape beats to death a marauding ape, and in an exaltation of victory, flings the bone-cum-weapon high into the air. The following shot is of a futuristic spaceship, controlled in large part by a computer, floating through space. Lest the meaning be misconstrued, the music accompanying the origin of tool use is from the portentous "Thus Spoke Zarathustra," while the spaceship floats on the gossamer wings of the "Blue Danube Waltz," written by Richard and Johann Strauss respectively. We need not follow Stanley Kubrick's metaphor for technology and intelligence in further detail. Although his gloomy prognostications certainly fit the distopian views of many scholars who ponder the relation between human nature and technology, they are almost certainly deficient in terms of contemporary theories of hominization (Bogin, 2001). It is sufficient for our purposes to note the widespread view that advances in human intelligence and the evolution of technology are intimately related.

TECHNOLOGY AND INTELLIGENCE RE-VIEWED

In seeking to contribute to a discussion of the relationship between technology and intelligence we immediately confront the difficulty that both concepts are conceived of in widely divergent terms by contemporary social scientists and the public alike. As the bulk of the articles in this volume indicate, the term *technology* evokes thoughts of computers, telecommunications networks, and spaceships, the technologies that occupy 99.9% of the story in the film *2001.* This view of technology fits well notions of intelligence that Neisser, Sternberg, and others have referred to as "academic intelligence," for example, the sorts of problem-solving skills that result in constructing computer networks and exploring outer space. It is a form of intelligence associated with modern schooling in which problems are generally formulated by others, well defined, have single correct answers, and single correct means of reaching those answers (Neisser, 1976). As a rule, this form of intelligence is treated as a biological property of individuals.

In our view, technology and intelligence understood in this manner are likely to underestimate what we believe to be an intimate, even incestuous, relationship between the two terms. To begin with, our view of technology leads us backward in time to the early evolution of *Homo sapiens* and such crude technologies as stone tools. This same view forces our attention to the fact that although tools may be considered constituents of technology, the concept of the tool itself needs to be reexamined, and the concept of technology broadened. Our perspective derives, in part, from the idea of technology that comes down to us from the Greeks: "A discourse or treatise on art or arts; the scientific study of the practical or industrial arts." Examples of early uses of the term in English indicate its range quite well (e.g., "His technology consists of weaving, cutting canoes, making crude weapons, and in some places practicing a crude metallurgy" (taken from an ethnographic description in the mid-19th century; Oxford English Dictionary). Essential to this broader notion of technology is that although tools are constituents of a technology, it is the way in which tools are deployed as part of a social practice that is crucial. As archaeologist Michael Schiffer puts its, the study of technology "must focus on behavior and artifacts in the context of activities" (Schiffer, 1992, p. x).

Our emphasis on technologies as forms of tool-mediated social practices also inclines us to adopt a broader notion of intelligence than that adopted in most contemporary theorizing on the subject. In its most general meaning, intelligence is better conceived of (following Piaget, 1952) as a process of adaptation to, and transformation of, the conditions of life. Important as it is to contemporary life, academic intelligence and the technological innovations it generates are not representative of life's adaptive endeavours—or as Binet and Simon noted, there is more to school than intelligence and more to life than school (Binet & Simon, 1916/1980, p. 256). In support of this perspective, a number of scholars have pointed out that in a great many situations, people must recognize or formulate problems that are of direct significance for their well being, are often poorly defined, require the acquisition of new information, and allow multiple routes to solution (Neisser, 1976; Sternberg et al., 2000). Hence, a theory of technology and intelligence from our perspective must take into account not only the means, but the conditions, of thought and the thinker, all of which have generally evolved in close interaction with each other (Semaw et al., 2003). A part must not be taken for the whole.

Artifacts: The Foundation Blocks of Technology

Thus far, we hope to have induced the reader to consider the possibility that there is an intimate relationship between technology and human intelligence, both conceived in unusually broad terms. Now we want to back up to consider the notion that technologies are constitutive of human nature in a deep sense that crosses the traditional lines between the mental and material, cognitive and noncognitive, and biology and culture. We begin putting the whole together by examining more closely the most fundamental element of any technology, the artifact.

The Dual Nature of Artifacts. In our usage, an artifact is an aspect of the material world that has been modified over the history of its incorporation into goal-directed human action. By virtue of the changes wrought in the processes of their creation and use, artifacts are *simultaneously ideal* *(conceptual) and material.* They are material in that they have been created by modifying physical material in the process of goal-directed human actions. They are ideal in that their material form has been shaped to fulfil the human intentions underpinning those earlier goals; these modified material forms exist in the present precisely because they successfully aided those human intentional goal-directed actions in the past, which is why they continue to be present for incorporation into human action.

The core of this idea was expressed by Dewey in the following terms: Tools and works of art, he wrote, "are simply prior natural things reshaped for the sake of entering effectively into some type of [human] behavior" (Dewey, 1916, p. 92).

The broad implications of the dual material–conceptual nature of artifacts were elaborated on by the Russian philosopher Evald Ilyenkov (1977, 1979), who based his approach on that of Marx and Hegel. As we have done, Ilyenkov and his followers emphasized that the form of an artifact is more than a purely physical form:

> Rather, in being created as an embodiment of purpose and incorporated into life activity in a certain way—being manufactured for a *reason* and put into *use*—the natural object acquires a significance. This significance is the "ideal form" of the object, a form that includes not a single atom of the tangible physical substance that possesses it. (Bakhurst, 1990, p. 182)

What is important to us is that this view asserts the primal unity of the material and the symbolic in human cognition. This starting point is important because it provides a way of avoiding dualistic approaches to the relation between the mental and the material in human life and overcoming Cartesian dualism in theories of thinking, which locate mind entirely inside the human brain.[1]

Kinds of Artifacts

Although they share defining features, artifacts differ from each other in a number of ways and are not haphazardly incorporated into human activity.

Differentiating Artifacts by Levels. The late American philosopher Marx Wartofsky proposed that artifacts can be usefully distinguished by levels. As examples of *primary artifacts* Wartofsky mentions axes, bowls, needles,

---

[1]Dewey believed that the tools and artifacts we call technological may be found on either side of what he argued was an extremely malleable and permeable membrane that separates the "internal" from the "external" with respect to the organism only in the loosest and most tentative senses (Hickman, 1990, p. 12).

clubs, etc. Their materiality is so manifest to us that the ideality built into their form is all but invisible. Whereas all human productive activity involves the use of primary artifacts, the modes of action and goals that accompany their use are in turn constituents of *secondary artifacts* (social forms of organizing action, relations of kinship), which enable the preservation and transmission of modes of action using primary artifacts. Although couched in somewhat different language, there are a great many suggestions about secondary artifacts as constituents of human activity. For example, anthropologist Roy D'Andrade suggested the term *cultural schemes* to refer to units that mediate entire sets of conceptual–material artifacts. In D'Andrade's terms:

> Typically such schemes portray simplified worlds, making the appropriateness of the terms that are based on them dependent on the degree to which these schemes fit the actual worlds of the objects being categorized. Such schemes portray not only the world of physical objects and events, but also more abstract worlds of social interaction, discourse, and even word meaning (1984, p. 93).

Psychologists such as Jerome Bruner (1990) and Katherine Nelson (1981) identify event schemas, embodied in narratives, as basic organizers of cognition. Referred to as *scripts* by Nelson, these generalized event schemes specify the people who participate in an event, the social roles that they play, the objects that are used during the event, the sequences of actions required, the goals to be attained, and so on. Nelson's account of scripted activity is similar in many ways to D'Andrade's suggestion that cultural schemas are the basic units of organized cognitive action.

Finally, Wartofsky identified special kinds of artifacts that he termed *tertiary artifacts.* These artifacts, he wrote, are ones in which "the forms of representation themselves come to constitute a 'world' (or 'worlds') of imaginative praxis" (Wartofsky, 1979, p. 207), allowing an arena for the playing out of broader intentions and affective needs.

Although each kind of artifact may be considered independently of the others, each, with its own mixtures of materiality and ideality arises from, and acts back on, the other. It is in this way that human beings bootstrap the means of their own cognition.

One of Wartofsky's main points is that environment is not a neutral term because it is changed by organisms and populations of organisms, and in the case of humans that transformation results from activity that includes artifacts. "Nature becomes transformed, not only in the direct practical way of becoming cultivated, or shaped into objects of use in the embodied artifacts we call tools . . . it becomes transformed as an object or arena of action, so that the forest or river is itself an 'artifact' in this ramified sense" (Wartofsky, 1979, p. 207).

In the same sense Ilyenkov presents the idea of nature as *idealized.* Meaning is embodied in the environment in which individuals are active. This view takes us toward a radical alternative to the dualism endemic in conceptualizations of human cognitive capacity, in which human physiology is realized only in an environment rich with the means of cognition. From this alternative perspective, intelligent activity arises as humans are able to orient themselves in the idealized environment that is the expression of nature in its human aspect. At each "level" of activity more is entailed than is initially the object of an activity.

Differentiating Artifacts by Function: Cognitive "Versus" Noncognitive Artifacts? Following the path laid out by Wartofsky, we are encouraged to differentiate artifacts by their levels, from those that mediate specific human actions to modes of action requiring the deployment and sequencing of many primary artifacts, to imaginative alternative worlds, to "anything which human beings create by the transformation of nature and of themselves: thus also language, forms of social organization and interaction, techniques of production, skills" (Wartofsky, 1979, p. xiii). Clearly, the exercise of intelligence is implicated in all forms of artifact-mediated human interaction.

Nor is this a uniquely modern insight. Even those who have focused rather narrowly on technology as primary artifacts, tools that amplify particular forms of human action, are likely to make the further claim that tools change not only actions directed outward on the world, but change the process of thought itself. For example, in the 17th century, Sir Francis Bacon, arguably one of the most important progenitors of contemporary science, declared:

> *nec manus, nisi intellectus, sibi permissus, mutlam valent: instrumentis et auxilibus res perfictur.* [The unassisted hand and understanding left to itself possess but little power.] Effects are produced by the means of instruments and aids, which the understanding requires no less than the hand; and as instruments either promote or regulate the motion of the hand, so those that are applied to the mind prompt or protect the understanding. (Bacon, 1620/1854, p. 345)

In the early 20th century, the French philosopher Henri Bergson spoke for many in this tradition when he wrote that:

> If we could rid ourselves of all pride, if, to define our species, we kept strictly to what the historic and prehistoric periods show us to be the constant characteristic of man and of intelligence, we should say not *Homo Sapiens* but *Homo Faber.* In short, *intelligence, considered in what seems to be its original feature, is the faculty of manufacturing artificial objects, especially tools for making tools, and of indefinitely varying the manufacture.* (Bergson, 1911/1983, p. 139)

Strikingly absent in these early statements of how human intelligence is linked to mediation of human action through tools, although present in Wartofsky's writings, is the idea that there is a category of artifacts that are expressly designed to influence some aspect of human thought. Lev Vygotsky referred to this category of artifacts as *psychological tools.* As examples of psychological tools he listed all kinds of symbolic cultural artifacts including not only linguistic signs and symbols, but counting schemes, mnemonic devices, diagrams, maps, all of which enable human beings to master psychological functions such as memory, perception, and attention "from the outside" (Wertsch, 1985).

In the early 1990s Donald Norman (who had not, so far as we know, encountered the ideas of Ilyenkov, Vygotsky, or Wartofsky) began to promote the idea of *cognitive artifacts* (Norman, 1991, 1993). Citing the general argument we have made that the creation and use of artifacts is central to human nature, Norman defined cognitive artifacts as "an artificial device designed to maintain, display, or operate upon information in order to serve a representational function" (Norman, 1991, p. 17).

The idea of a representation is not defined precisely, but the idea is clear enough from both the remarks Norman makes about representations and the examples he provides. Pooling this information (Norman, 1993, pp. 49–51 and 1991, p. 25 ff) we can say that:

- A representation is a set of symbols that substitutes for the real event.
- Once we have ideas represented by representations, the physical world is no longer relevant.
- Representations are abstractions, so good representations are those that abstract the essential elements of the event.
- The critical trick is to get the abstractions right, to represent the important aspects and not the unimportant. This allows everyone to concentrate on the essentials without distraction from irrelevancies.
- Representations are important because they allow us to work with events and things absent in space and time, or for that matter, events and things that never existed—imaginary objects and concepts.
- A person is a system with an active, internal representation.

At many points in his discussion, Norman makes clear that cognitive artifacts are *extrinsic* to human thought, external complements to naturally occurring internal representations. First, he states this view quite directly as a general premise for his treatment of artifacts, asserting that he wants to "emphasize the information-processing role played by physical artifacts upon the cognition of the individual" (1991, p. 18).

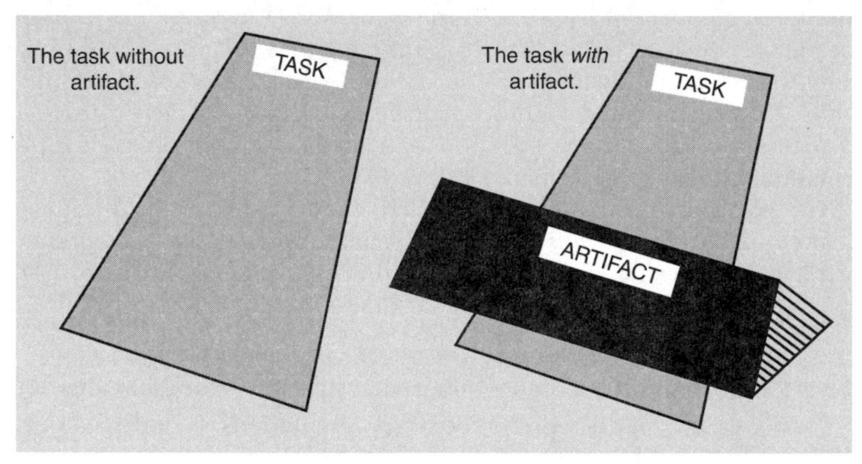

Fig 10.1. Donald Norman's two views on the relation of artifacts to cognition. The left-hand diagrams in both halves of the figure represent the individual person's point of view, whereas the right-hand diagrams represent the system's point of view.

Norman elaborates on the separation between cognitive artifacts and "natural" human thought by offering a contrast between two views of artifacts, a *system view* and a *personal view* (see Fig. 10.1).

To use an example that Norman himself proposes, let's assume that the artifact in question is written language, a list of things that the person seeks to remember—say, an airplane pilot reading a checklist in preparation for a flight. From a system view, he argues, it appears that one is dealing with a total structure inclusive of person, artifact, and task. The artifact appears intrinsic to the act of remembering. But from the perspective of the individual person, the artifact has simply changed the task. In fact, reading the list has itself become a task. Norman summarizes the situation as follows:

> Every artifact has both a system and a personal point of view, and they are often very different in appearance. From the system view, the artifact appears to expand some functional capacity of the task performer. From the personal point of view, the artifact has replaced the original task with a different task, one that may have radically different cognitive requirements and use radically different cognitive capacities than the original task. (1991, p. 22)

At first blush, it seems difficult to avoid the conclusion that what Norman refers to as the "personal point of view" is simply a confusion. For the airplane pilot, a written list mediates action. The goal of the action, with or without the list, is to have a safe flight. However, Norman is here drawing on a long tradition in psychology that defines a cognitive task as a goal and the constraints on achieving it. From this perspective, any change in the means by which the goal is achieved *ipso facto* changes the nature of the task.

We can see a certain heuristic value to making this strong distinction between internal and external representations and the implied distinction between cognitive and noncognitive artifacts. Artifacts that partake of the cognitive, in this view, should be studied in terms of the kinds of representations they can encompass. A voice recognition device, for example, would be an excellent example of a cognitive artifact because of the enormous amount of representational information it contains. The analyst's task becomes one of figuring out the most natural interface between the input–output capacities of the device and the (internal) representational state of the user. Or, as Norman puts it:

> We can conceptualize the artifact and its interface in this way. A person is a system with an active, internal representation. For an artifact to be usable, the surface representation must correspond to something that is interpretable by the person, and the operations required to modify the information within the artifact must be performable by the user. The interface serves to transform the properties of the artifact's representational system to those that match the properties of the person. (1991, p. 22)

When all is said and done, Norman's use of the notion of cognitive artifact enables him to argue that cognitive artifacts "serve human cognition." The idea that all artifacts, like all human action (including the kind of action we refer to as thinking), are at once ideal and material is lost. As a consequence, Norman speaks of cognition being distributed among humans by virtue of shared action involving artifacts, but cannot conceive of the possibility that cognition is a mediated interaction, always involving other people and the artifact-saturated environment.

Intriguingly, Wartofsky also used the term *cognitive artifact,* commenting at one point that "we create cognitive artifacts which not only go beyond the biologically evolved and genetically inherited modes of perceptual and cognitive activity, but which radically alter the very nature of learning and which demarcate human knowledge from animal intelligence" (Wartofsky, 1979, p. xv).

But for Wartofsky, cognitive artifacts such as representations are not *what* we perceive. Rather, they are *the means by which* we perceive real objects. This distinction, though apparently trivial, is key to appreciating the active and practical nature as well as the external (socio) genesis of our cognitive capacities (more about this in the following). Wartofsky speaks of our faculty of perception as the result of activity rather than as a capacity: "I take perception itself to be a mode of outward action or praxis. In this sense, it is perceptual activity *in* the world, and *of* a world as it is transformed by such activity" (Wartofsky 1979, p. 194). But also coupled to the idea of cognition as activity is a conception of knowledge that rejects a given on which our "theory-dependent observation" selects features of nature. Rather what is there or given in nature is already a product of material activity. The form of production and reproduction of the human species takes place with the use of tools/artifacts in the sense that human activity is goal-oriented, transforming the environment to fit our purposes rather than merely inhabiting what is made available at any point by nature.

The Functional Structure of Artifact-Mediated Action. Regardless of the properties they attribute to artifacts, those who claim a strong link between human technologies and human intelligence believe that tools/technologies mediate human action. In the Russian cultural–historical tradition on which we draw, the relation of artifacts to human action is likely to be depicted as a triangle representing the structural relation of the individual to environment that arises *pari passu* with artifact use (see Fig. 10.2;

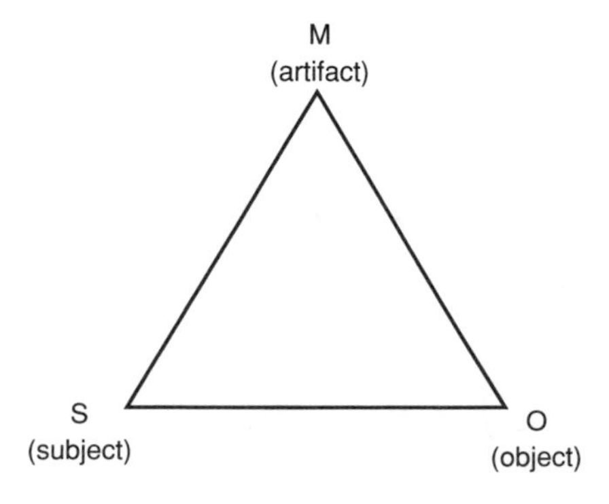

Fig 10.2. The basic mediational triangle in which subject and object are not seen only as "directly" connected but simultaneously as "indirectly" connected through a medium constituted of artifacts (culture).

Vygotsky, 1929, 1978). Simplifying this view for purposes of explication, we can say that the functions termed *natural* or *unmediated* are those along the base of the triangle; the *cultural* (*mediated*) functions are those where the relation between subject and environment (subject and object, response and stimulus, etc.) are linked through the vertex of the triangle (artifact).

There is some temptation when viewing this triangle to think that when artifacts are incorporated into human action, thought follows a path through the top line of the triangle, for example, that it *runs through* the mediator. However, the emergence of mediated action does not mean that the mediated path *replaces* the natural one, just as the appearance of culture in phylogeny does not replace phylogeny by culture. One does not cease to stand on the ground and look at the tree when one picks up an axe to chop the tree down; rather, the incorporation of tools into the activity creates a new structural relation in which the cultural (mediated) and natural (unmediated) routes operate synergistically; through active attempts to appropriate their surroundings to their own goals, people incorporate auxiliary means (including, very significantly, other people) into their actions, giving rise to the distinctive, triadic relationship of subject–medium–object.

Even this basic notion that human thought is the emergent consequence of intermingling of direct/natural/phylogenetic and indirect/cultural/historical aspects of experience is sufficient to bring to the fore the special quality of human thought referred to as the duality of human consciousness. As the American anthropologist Leslie White wrote[2]:

> An axe has a subjective component; it would be meaningless without a concept and an attitude. On the other hand, a concept or attitude would be meaningless without overt expression, in behavior or speech (which is a form of behavior). Every cultural element, every cultural trait, therefore, has a subjective and an objective aspect (1959, p. 236).

A great deal more can be said about this basic conception of artifact-mediated human action and the peculiar form of consciousness to which it gives rise (Cole, 1996). However, artifacts and artifact-mediated individual human action are only a *starting* point for developing the needed conceptual tools for thinking about technology and intelligence. Neither artifacts nor actions exist in isolation. Rather, they are interwoven with each other and the social worlds of the human beings they mediate to form the vast networks of interconnections known as human culture (Ellul, 1980; Latour, 1993).

---

[2]Richard Barrett (1989) provides a useful discussion of White's symbolic/mediational views in relation to his better known views concerning materialist evolutionism.

FROM ARTIFACTS TO CULTURE

Implicit in our discussion thus far, and stated directly by White in the passage quoted, is an implied but unexplicated claim that there is a close relation between the nature of artifacts and the nature of culture. It is time to make that linkage clear.

In its most general sense, the term *culture* is used to refer to the socially inherited body of past human accomplishments that serves as the resources for the current life of a social group ordinarily thought of as the inhabitants of a country or region (D'Andrade, 1996). In trying to specify more carefully the notion of *culture-as-social-inheritance,* anthropologists have historically tended to employ the same dichotomy to culture that we have sought to supersede with respect to the concept of artifact. As Roy D'Andrade has noted, during the first half of this century, the notion of culture as something "superorganic" and material dominated anthropological thinking, but as a consequence of the cognitive revolution in the social sciences, the pendulum shifted, so that for several decades, a *culture-as-knowledge* view has reigned. This latter view is most closely associated with the work of Ward Goodenough, for whom culture consists of "what one needs to know to participate acceptably as a member in a society's affairs" (Goodenough, 1994, p. 265). This knowledge is acquired through learning and, consequently, is a mental phenomenon. As Goodenough put it:

> Material objects people create are not in and of themselves things they learn. . . . What they learn are the necessary percepts, concepts, recipes, and skill—the things they need to know in order to make things that will meet the standards of their fellows. (p. 50)

From this perspective, culture has little do to with artifacts, which are considered a part of material culture, whereas the real stuff of culture is profoundly subjective. It is in people's minds, the mental products of the social heritage.

However, just as we and other psychologists are seeking to transcend this ideal versus material culture dichotomy, so too have anthropologists. For example, in an oft-quoted passage, Clifford Geertz (1973) wrote that his view of culture begins with the assumption that:

> Human thought is basically both social and public—that its natural habitat is the house yard, the market place, and the town square. Thinking consists not of "happenings in the head" (though happenings there and elsewhere are necessary for it to occur) but of trafficking in . . . significant symbols—words for the most part but also gestures, drawings, musical sounds, mechanical devices like clocks. (p. 45)

Geertz, coming at the problem from a quite different direction than we have taken, provides an escape from the ideal–material dichotomy with respect to culture that dovetails perfectly with the idea that human beings live in an environment transformed by the artifacts of prior generations. The basic function of these artifacts is to coordinate human beings with the physical world and each other; in the aggregate, culture is then seen as the species-specific *medium* of human development.

D'Andrade (1986) made this point when he said that "Material culture—tables and chairs, buildings and cities—is the reification of human ideas in a solid medium" (p. 22). As a consequence of the dual conceptual–material nature of the systems of artifacts that are the cultural medium of their existence, human beings live in a double world, simultaneously natural and artificial.

Geertz's reference to the house yard, the market place, and the town square remind us that it is insufficient to think of artifacts as all of a piece or haphazardly strewn around the environment. Rather, they are better considered as constituents of cultural practices, each of which aggregates artifacts into different kinds of technologies for dealing with the world at hand.

Arranging for the Acquisition of Technologies

The views of tool use as both amplifier of human action and transformative of human mind, and that technology, taken as a whole, constitutes the special environment of human life, take on even broader significance when they are combined with a theory of human development. Such a theory assumes that cognitive development depends crucially on the ways in which adults arrange the environment so that as children interact with more mature members of the social group, they simultaneously acquire the cultural toolkit (ensemble of technologies) that is the group's social inheritance. This idea, which can be traced back to Janet (see Valsiner, 2000), has received its most influential formulation in what Vygotsky referred to as "the general law of cultural development":

> Any function in children's cultural development appears twice, or on two planes. First it appears on the social plane and then on the psychological plane. First it appears between people as an interpsychological category and then within the individual child as an intra-psychological category... but it goes without saying that internalization transforms the process itself and changes its structure and function. Social relations or relations among people genetically underlie all higher functions and their relationships. (Vygotsky, 1981, p. 163)

The idea that interpsychological processes (transactions between people) precede intrapsychological processes (complex mental processes in the child's mind) appears counterintuitive when mind is understood as an inbuilt individual capacity that matures on an invariant time schedule. However, the view that interpsychological processes precede intrapsychological processes is a natural conclusion if one starts from the assumption that older members of the community are bearers of the intellectual tool kit of the social group. That tool kit is essential both to the group's survival and to the development of mind, so that transactions between adults and children are the means for the individual's appropriation of the knowledge essential to the development of the mind. This latter view, which we adopt in this chapter, can be summarized by saying that all means of social behavior (technologies) are social in their essence (and in the dynamics of their origin and change) so that the structure and development of human intelligence emerges through culturally mediated, historically developing, practical activity. Furthermore, this statement applies equally to the phylogeny and ontogeny of human intelligence, broadly understood.

THE PHYLOGENETIC INTERWEAVING OF ARTIFACTS, CULTURE, AND THE HUMAN BRAIN

Even if the reader accepts our claim about the priority of the social group in the development of specifically human psychological abilities, the idea that human phylogeny also involves culturally mediated, historically developing, practical activity may seem a bit odd.

However, because artifacts aggregated into technologies (for killing and cutting up large animals for food, for transforming their skin into clothing, and for sources of shelter, etc.) that have been present for perhaps 2.5 million years prior to the emergence of *Homo sapiens*, it is not appropriate to focus on technology and intelligence without including human biological as well as technological/cultural evolution. The human brain and body coevolved over a long period of time with our species' increasingly complex cultural environment (Plotkin, 2003; Quartz & Sejnowski, 2002; Semaw et al., 2003).

When Clifford Geertz (1973) examined the mounting evidence that the human body, and most especially the human brain, underwent a long coevolution with the basic ability to create and use artifacts he was led to conclude that:

> Man's nervous system does not merely enable him to acquire culture, it positively demands that he do so if it is going to function at all. Rather than culture acting only to supplement, develop, and extend organically based capacities logically and genetically prior to it, it would seem to be ingredient to those capacities themselves. A cultureless human being would probably turn out to be not an intrinsically talented, though unfulfilled ape, but a wholly mindless and consequently unworkable monstrosity. (p. 68)

Despite 30 years of intensive research on this issue and all of the controversies one would expect given the many remaining gaps in the evolutionary record, Geertz's main point appears secure. The human brain of modern *Homo sapiens* is several times larger and more complex than the brain of *Homo habilus*, among whom the first rudimentary tools were discovered. Moreover, that growth took place in an environment that was increasingly influenced by the products of (proto)human activity. In short, the human brain evolved in an environment increasingly modified by human culture, such that interaction through culture/technology became an essential design feature of *both* human biology *and* the human life world. As neuroscientists Steven Quartz and Terrence Sejnowski summarize, "culture plays a central role in the development of the prefrontal cortex.... [so that] Culture, then, contains part of the developmental program that works with genes to build the brain that underlies who you are" (Quartz & Sejnowski, 2002, p. 58). They emphasize, especially, the fact that the prefrontal cortex, which is the latest brain structure to develop in both phylogeny and ontogeny, and which is central to planning functions and complex social interaction, depends crucially on culture for its development.

Quartz and Sejnowski develop a broad view of culture as "groupwide practices that are passed down from one generation to the next" (Quartz & Sejnowski, 2002, p. 82) and note that traces of culture can be found in our near phylogenetic neighbors. Symptomatically, they adopt a corresponding broad view of intelligence that encompasses both its academic and everyday features, commenting that, "Not only is intelligence a complex strand of social, emotional, intellectual, and motivational brain systems, but the central role of culture in our mental life reveals that intelligence isn't just inside the head" (2002, p. 233).

Quartz and Sejnowski mention neither the notions of technology nor of tool in their fascinating presentation of what they refer to as "cultural biology." But they do make a comment that provides a natural and productive bridge between their approach and that which we adopt when they comment that, "The artifacts of human culture are unlike anything ever seen in the three-billion-year history of life on earth" (Quartz & Sejnowski, 2002, p. 67).

In order to make progress in fleshing out these ideas, we believe it is important to note that the invocation of culture with respect to near-phylogenetic cousins and progenitors refers to practices with no reference to artifacts, whereas it is the artifacts of human culture that appear to be the locus of inter-phylogenetic discontinuity.

In this connection, the example of archeologists that was used by philosophers in the interwar period is pertinent. In understanding objects such as a lost city or particular artifacts they pointed out, the knowledge of their natural material was of limited use. What was critical was to know the reasons and purpose behind the object. Foster quotes an archeologist who wrote, "We found cuttings in the rocks which puzzled us for a long time till [we discovered] they were wine presses." He continues, "This discovery was not a detection by any of the senses of sensible qualities which had hitherto" not been known, it was the discovery of the purpose for which the cuttings had been made (Foster, 1934, p. 460; see also Schiffer, 1992). Drawing on a large body of theory and research from that branch of cultural psychology referred to as cultural–historical-activity theory, which allows us to link artifacts and practices to the notion of technology, we believe we can establish the complementarity of Quartz and Sejnowski's approach, stemming from their deep knowledge of neuroscience, with an approach that begins with scholarship of the study of human development in its cultural and historical contexts.

This long-term, phylogenetic perspective is also important to keep in mind when considering the ontogeny of children, for it reminds us that causal influences do not run unidirectionally from biology to culture. Rather, human beings are hybrids of phylogenetic, cultural–historical, and ontogenetic sources. Activity-dependent influences, no less than activity-expectant processes, shape the development of the human brain (Cole, 1996).

Nature Through Nurture: Working Through an Example

The position we have developed in this chapter strongly urges us to keep in mind the bi-directional influences between culture and biology that no longer appear as polar opposites, but as intertwined aspects of human nature. We end our discussion using a phrase that is the title of a recent book by Henry Plotkin, well known for his writings on Darwinism (Plotkin, 2003). Like Quartz and Sejnowski, as well as ourselves, Plotkin argues for a view of culture and the social origins of higher human psychological functions consistent with the ideas of Vygotsky and our view of the duality of artifacts. But, like many who are discovering this mode of thinking about technology and human nature, Plotkin's discussion remains at a relatively general level that needs filling in with concrete, well-worked-out examples that range across phylogeny, cultural history, ontogeny, and microgenesis. In the spirit of this effort, we provide one such example, for which there is more than the usual amount of evidence concerning brain–technology–ontogeny relations, although there is still much to be worked out.

The case of the use of the abacus by Japanese school children and adults provides an illustration of how thoroughly the historical processes involved in the development of a tool's use becomes incorporated into a culture-specific technology while simultaneously becoming a part of human nature.

With respect to phylogeny, the most that we can say is that there is currently a good deal of evidence for at least rudimentary arithmetic abilities in nonhuman primates (Boysen & Hallberg, 2000), but there is no known case of the use of artifacts in this process, let alone an artifact as complex as an abacus. The abacus, which traces its origins back several thousand years to Sumer in the fertile crescent, was introduced into Japan from China, where it appears to come into use in the 14th century. For many centuries the Japanese have used the abacus (referred to as *sokoban* in Japanese) as a basic tool for mathematical calculations (Ifrah, 2000).

Since its introduction, this tool has spread *outward* to form around itself a set of social practices that render it a technology while simultaneously burrowing *inward* to become a mental tool with a specific localization in the brain for those who become expert in its use. Giyoo Hatano and his colleagues, who have been leaders in studying the psychological consequences of this technology, report that use of the abacus is introduced into the elementary school curriculum around the third grade, following the introduction of paper-and-pencil algorithmic techniques in the first and second grade (Hatano, 1997). But involvement in using the abacus is not restricted to the formal school curriculum. Rather, there are special after-school schools (juku) that specialize in teaching use of the abacus and especially the skill of making calculations using a "mental abacus," an image of the real thing, which allows experts to carry out very large calculations in their heads (although movements of their fingers often accompany such calculations). There are also clubs that form to permit children and adults, who often practice using an abacus two or more hours a day, to engage in tournaments, much in the spirit of American intercollegiate sports. There is a national organization that has created a standardized examination with 10 grades of mastery. In 1971 more than 2 million Japanese had taken this examination.

Considerable research indicates that achieving high levels of skill in the use of the mental abacus is associated with improved mathematical performance that involves much more than bare calculation (summarized in Hatano, 1997). Moreover, current research has begun to direct itself toward understanding the brain basis of high levels of abacus training (Hanakawa, Okada, Fukuyama, & Shibasaki, 2003; Tanaka, Michimata, Kaminaga, Honda, & Sadato, 2002). Whether tested for digit memory or mental arithmetic, Functional Magnetic Resonance Imaging (fMRI) recordings of abacus experts engaged in such tasks show right hemisphere activation of the parietal area and other structures related to spatial processing. The fMRI activity in nonexperts engaged in such tasks is in the left hemisphere, including Broca's area, indicating that they are solving the task by language-mediated, temporally sequential processing. When compared to being engaged in verbal tasks, experts and nonexperts display the same forms of left hemisphere-dominated fMRI activity. Although a great deal more research is needed, the case of the abacus illustrates the way in which psychological tools incorporated into cultural practices constitute those practices as technologies and that this experience reacts back on the human brain. Nurture becomes nature.

This example also points to the kind of interdisciplinary work that will be needed to carry the study of technology and human nature/intelligence forward in the years to come. What is called for are interdisciplinary teams, ideally, but not necessarily, located in the same institutions, who can help each other span the enormous range of expertises necessary to encompass phylogenetic, cultural–historical, ontogenetic, and microgenetic processes (including online brain imaging), bringing them together in single research efforts. It is not an easy goal to achieve. At least we now have a better grip on what the development of more powerful theorizing about technology and human nature requires.

REFERENCES

Bakhurst, D. (1990). *Consciousness and revolution in Soviet philosophy: From the Bolsheviks to Evald Ilyenkov.* Cambridge, UK: Cambridge University Press.

Bacon, F. (1854). The aphorisms. In B. Montague (Ed. and Trans.), *The interpretation of nature: The works* (Vol. 3, p. 345). Philadelphia: Parry & Macmillan. (Original work published 1620)

Barrett, R. A. (1989). The paradoxical anthropology of Leslie White. *American Anthropologist, 91,* 986–999.

Bergson, H. (1983). *Creative evolution.* New York: Holt. (Original work published 1911)

Binet, A., & Simon, T. (1980). *The development of intelligence in children.* Nashville, TN: Williams. (Original work published 1916)

Bogin, B. (2001). *The growth of humanity.* New York: Wiley Liss.

Boysen, S. T., & Hallberg, K. I. (2000). Primate numerical competence: Contributions toward understanding nonhuman cognition. *Cognitive Science, 24*(3), 423–443.

Bruner, J. (1990). *Acts of meaning.* Cambridge, MA: Harvard University Press.

Cole, M. (1996). *Cultural psychology: A once and future discipline.* Cambridge, MA: Harvard University Press.

D'Andrade, R. (1984). Cultural meaning systems. In R. A. Shweder & R. A. Le Vine (Eds.), *Culture theory: Essays on mind, self and emotion.* New York: Cambridge University Press.

D'Andrade, R. (1986). Three scientific world views and the covering law model. In D. Fiske & R. A. Shweder (Eds.), *Metatheory in the social sciences.* Chicago: University of Chicago Press.

D'Andrade, R. (1996). Culture. In A. Kuper & J. Kuper (Eds.), *Social science encyclopedia.* (2nd ed., pp. 161–163). London: Routledge.

Dewey, J. (1916). *Human nature and experience.* New York: Holt.

Ellul, J. (1980). *The technological system.* New York: Continuum.

Foster, M. B. (1934). The Christian doctrine of creation and the rise of modern natural science, *Mind, New Series, 43*(172), 460.

Geertz, C. (1973). *The interpretation of cultures.* New York: Basic Books.

Goodenough, W. H. (1994). Toward a working theory of culture. In R. Borovsky (Ed.), *Assessing cultural anthropology* New York: McGraw-Hill. (pp. 262–273).

Hanakawa, T., Okada, T., Fukuyama, H., & Shibasaki, H. (2003). Neural correlates underlying mental calculations in abacus experts: A functional magnetic resonance imaging study. *NeuorImage, 19,* 296–307.

Hatano, G. (1997). Learning arithmetic with an abacus. In T. Nunes & P. Bryant (Eds.), *Learning and teaching mathematics: An international perspective* (pp. 209–232). Hove, UK: Psychology Press.

Hickman, L. (1990). *Dewey's pragmatic technology.* Bloomington: Indiana. University Press.

Ifrah, G. (2000). *The universal history of computing.* New York: Wiley.

Ilyenkov, E. V. (1977). The problem of the ideal. In *Philosophy in the USSR: Problems of dialectical materialism* (pp. 71–99). Moscow: Progress.

Ilyenkov, E. V. (1979). Problema ideal'nogo [The problem of the ideal]. *Voprosy filosofi* [Questions of Philosophy], *6,* 145–158 and *7,* 126–140.

Latour, B. (1993). *We never have been modern.* Cambridge, MA: Harvard University Press.

Neisser, U. (1976). General, academic, and artificial intelligence. In L. Resnick (Ed.), *Human intelligence: Perspectives on its theory and measurement* (pp. 179–189). Norwood, NJ: Ablex.

Nelson, K. (1981). Cognition in a script framework. In J. H. Flavell & L. Ross (Eds.), *Social congnitive development.* Cambridge: Cambridge University Press.

Norman, D. A. (1991). Congnitive artifacts. In J. M. Carroll (Ed.), *Designing interaction: Psychology at the human-computer interface* (pp. 17–38). New York, NY, US: Cambridge University Press.

Norman, D. A. (1993). *Things that make us smart: defending human attributes in the age of the machine. Reading,* MA: Addison-Wesley Pub. Co.

Piaget, J. (1952). *The origins of intelligence in the child.* New York: International Universities Press Plotkin, 2003.

Plotkin, H. (2003). *The imagined world made real: Towards a natural science of culture.* New Brunswick, NJ: Rutgers University Press.

Quartz, S. R., & Sejnowski, T. J. (2002). *Liars, lovers, and heroes: what the new brain science reveals about how we become who we are.* New York, NY: William Morrow.

Schiffer, M. B. (1992). *Technological persperctives on behavioral change.* Tucson, AZ: University of Arizona Press.

Semaw, S., Rogers, M. J., Quade, J., Renne, P. R., Butler, R. F., Dominguez-Rodrigo, M., Stout, D., Hart, W. S., Pickering, T., & Simpson, S. W. (2003). 2.6-Milliom-year-old stone tools and associated bones from OGS-6 and OGS-7, Gona, Afar, Ethiopia. *Journal of Human Evolution. 45*(2), 169–77.

Sternberg, R. J., Forsythe, G. B., Hedlund, J., Horvath, J., Snook, S., Williams, W. M., Wagner, R. K., & Grigorenko, E. L. (2000). *Practical intelligence in everyday life.* New York: Cambridge University Press.

Tanaka, S., Michimata, C., Kaminaga, T., Honda, M., & Sadato, N. (2002). Superior digit memory of abacus experts: an event-related functional MRI study. *Neuroreport, 13*(17), 2187–91.

Valsiner, J. (2000). *The guided mind.* Cambridge, MA: Harvard University Press.

Vygotsky, L. S. (1929). The problem of the cultural development of the child, II. *Journal of Genetic Psychology, 36,* 414–434.

Vygotsky, L. S. (1978). *Mind in society.* Cambridge, MA: Harvard University Press.

Vygotsky, L. S. (1981). *The genesis of higher psychological functions.* In J. V. Wertsch (Ed.). The concept of activity in soviet psychology. Armonk, NY: M. E. Sharpe.

Wartofsky, M. (1979). *Models.* Dordrecht, The Netherlands: Reidel.

Wertsch, J. (1985). *Vygotsky and the social formation of mind.* Cambridge, MA: Harvard University Press.

White, L. (1959). The concept of culture. *American Anthropologist, 61,* 227–251.

Author Index

Numbers in *italics* indicate pages with complete bibliographic information.

T

U

V

W

Y

Z

Subject Index

Note: *f* indicates figure and *t* indicates table.

D

K

L

M

U

V